Engineering Flow
and Heat Exchange

The Plenum Chemical Engineering Series

Series Editor
Dan Luss, *University of Houston, Houston, Texas*

ENGINEERING FLOW AND HEAT EXCHANGE
Octave Levenspiel

COAL COMBUSTION AND GASIFICATION
L. Douglas Smoot and Philip J. Smith

Engineering Flow and Heat Exchange

Octave Levenspiel

Oregon State University
Corvallis, Oregon

10ᵗʰ April 1986.

Plenum Press • New York and London

Library of Congress Cataloging in Publication Data

Levenspiel, Octave.
    Engineering flow and heat exchange.

    (The Plenum chemical engineering series)
    Bibliography: p.
    Includes index.
    1. Fluid dynamics. 2. Heat — Transmission. I. Title. II. Series.
TA357.L48   1984   620.1'064   84-13448
    ISBN 0-306-41599-2

©1984 Plenum Press, New York
A Division of Plenum Publishing Corporation
233 Spring Street, New York, N.Y. 10013

Printed in the United States of America

PREFACE

This volume presents an overview of fluid flow and heat exchange.

In the broad sense, fluids are materials which are able to flow under the right conditions. These include all sorts of things: pipeline gases, coal slurries, toothpaste, gases in high-vacuum systems, metallic gold, soups and paints, and, of course, air and water. These materials are very different types of fluids, and so it is important to know the different classifications of fluids, how each is to be analyzed (and these methods are quite different), and where a particular fluid fits into this broad picture.

This book treats fluids in this broad sense including flows in packed beds and fluidized beds. Naturally, in so small a volume, we do not go deeply into the study of any particular type of flow, however we do show how to make a start with each. We avoid supersonic flow and the complex subject of multiphase flow where each of the phases must be treated separately.

The approach here differs from most introductory books on fluids which focus on the Newtonian fluid and treat it thoroughly, to the exclusion of all else. I feel that the student engineer or technologist preparing for the real world should be introduced to these other topics.

Introductory heat transfer books are devoted primarily to the study of the basic rate phenomena of conduction, convection, and radiation, showing how to evaluate "*h*," "*U*," and "*k*" for this and that geometry and situation. Again, this book's approach is different. We rapidly summarize the basic equations of heat transfer, including the numerous correlations for "*h*". Then we go straight to the problem of how to get heat from here to there and from one stream to another.

The recuperator (or through-the-wall exchanger), the direct contact exchanger, the heat storing accumulator (or regenerator), and the exchanger which uses a third go-between stream—these are distinctly different ways of transferring heat from one stream to another, and this is what we concentrate on. It is surprising how much creativity may be needed to develop a good design for the transfer of heat from a stream of hot solid particles to a stream of cold solid particles. The flavor of this presentation of heat exchange is that of Kern's unique book; certainly simpler, but at the same time broader in approach.

Wrestling with problems is the key to learning and each of the chapters has illustrative examples and a number of practice problems. Teaching and learning should be interesting so I have included a wide variety of problems, some whimsical, others directly from industrial applications. Usually the information given in these practice problems has been designed so as to fall on unique points on the design charts, making it easy for the student and also for the instructor who is checking the details of a student's solution.

I think that this book will interest the practicing engineer or technologist who wants a broad picture of the subject, or on having a particular problem to solve, wants to know what approach to take.

In the university it could well form the basis for an undergraduate course in engineering or applied fluids and heat transfer, after the principles have been introduced in a basic engineering course such as transport phenomena. At present, such a course is rarely taught; however, I feel it should be an integral part of the curriculum, at least for the chemical engineer and the food technologist. Who knows, some day it may be.

Finally, my thanks to supertypists Laurie Campbell, Vi Campbell, and Nancy Platz for their boundless patience with this mid-space writer, to Richard Turton, who coaxed our idiot computer into drawing charts for this book, and to Eric Swenson, who so kindly consented to put his skilled hand to the creation of drawing and sketch to enliven and complement the text.

OCTAVE LEVENSPIEL

Corvallis, Oregon

CONTENTS

NOMENCLATURE

a	specific surface, surface of solid/volume of vessel [m^{-1}]
A	area normal to flow, exterior surface area of a particle, area of exchanger [m^2]
A_t	cross-sectional area of flow channel [m^2]
Ar	Archimedes number, for fluidized beds [−]; see equation (7.4) and Appendix T
Bi	Biot number, for heat transfer [−]; see equation (11.4) and Appendix T
c	speed of sound in the fluid [m/s]; see equation (3.2)
C_D	drag coefficients for falling particles [−]; see equation (8.2) and Appendix T
C_g, C_l, C_s	specific heat of gas, liquid, or solid at constant pressure [J/kg K]
C_p	specific heat at constant pressure [J/kg K]; see Appendices P and U
C_v	specific heat at constant volume [J/kg K]
C_{12}	conductance for flow between points 1 and 2 in a flow channel [m^3/s]; see equation (4.3)
d	diameter [m]
d_e	equivalent diameter of noncircular channel [m]; see equation (2.16) or (9.15)

d_p	characteristic diameter of particles to use in flow problems [m]; see equation (6.3)
d_{scr}	screen diameter of particles [m]; see discussion in text between equations (6.3) and (6.4)
d_{sph}	equivalent spherical diameter of particles [m]; see equation (6.2)
f_D	Darcy friction factor, for flow in pipes [−]; see text after equation (2.5)
f_f	friction factor, for flow in packed beds [−]; see equation (6.10) and Appendix T
f_F	Fanning friction factor, for flow in pipes [−]; see equation 2.1, Figs. 2.4 and 2.5, and Appendix T
F	force [N]
Fo	Fourier number, for unsteady state heat conduction [−], see equation (11.2) and Appendix T
$F_{12}, F'_{12}, \overline{F}_{12}, \mathscr{F}_{12}$	various view factors for radiation between two surfaces, fraction of radiation leaving surface 1 which is intercepted by surface 2 [−]; see equations (9.74), (9.79), (9.81), and (9.83)
$\mathscr{F}$	efficiency factor for shell-and-tube heat exchangers [−]; see equation (13.17a)
F_d	drag force on a falling particle [N]; see equation (8.1)
ΣF	lost mechanical work of a flowing fluid due to friction [J/kg]; see equation (1.5)
g	acceleration of gravity, about 9.8 m/s^2 at sea level [m/s^2]
$g_c = 1 \text{ kg m/s}^2 \text{ N}$	conversion factor needed to get a consistent system of units; see Appendix E for additional values
$G = u\rho = G_0/\varepsilon$	mass velocity of flowing fluid, that based on the mean cross-sectional area available for the flowing fluid in the packed bed [kg/m^2 open · s]
$G_0 = u_0\rho = G\varepsilon$	mass velocity of flowing fluid, that based on the total cross-sectional area of packed bed [kg/m^2 bed · s]
Gr	Grashof number, for natural convection [−]; see text above equation (11.31) and Appendix T
h	heat transfer coefficient, for convection [W/m^2 K]; see text above equation (9.11)
h_L	head loss of fluid resulting from frictional effects [m]; see equation (1.6) and the figure below equation (2.2)
H	enthalpy [J/kg]
He	Hedstrom number, for flow of Bingham plastics [−]; see equation (5.8) and Appendix T
i.d.	inside diameter [m]
k	thermal conductivity [W/m K]; see equation (9.1) and Appendices O and V
$k = C_p/C_v$	ratio of specific heats of fluid [−]; $k \cong 1.67$ for monotonic gases; $k \cong 1.40$ for diatomic gases; $k \cong 1.32$ for triatomic gases; $k \cong 1$ for liquids

K fluid consistency index of power law fluids and general plastics, a measure of viscosity $[kg/m\,s^{2-n}]$; see equations (5.3) and (5.4)

$KE = u^2/2g_c$ shorthand notation for the kinetic energy of the flowing fluid $[J/kg]$

Kn   Knudsen number, for molecular flow $[-]$; see beginning of Chapter 4 and see Appendix T

l length or distance $[m]$

L length of flow channel or vessel $[m]$

L, L_p characteristic length of particle $[m]$; see equation (11.3) and text after equation (15.10)

m mass of particle $[kg]$

$\dot{m}$ mass flow rate $[kg/s]$

M number of standard deviations; see equation (15.10)

$Ma = u/c$ Mach number, for compressible flow of gas $[-]$; see equations (3.1) and (3.2)

(mfp) mean free path of molecules $[m]$; see Chapter 4

(mw) molecular weight $[kg/mol]$; see Appendix J; $(mw) = 0.0289$ kg/mol, for air

n flow behavior index for power law fluids and general plastics $[-]$; see equations (5.3) and (5.4)

$\dot{n}$ molar flow rate $[mol/s]$

N number of stages in a multistage heat exchanger $[-]$; see Chapter 14

$N = 4f_F L/d$ pipe resistance term $[-]$; see equation (3.7)

N rotational rate of a bob of a rotary viscometer $[s^{-1}]$; see equation (5.15)

NNs   shorthand notation for non-Newtonian fluids

$NTU = UA/\dot{M}C$ number of transfer units $[-]$; see Fig. 11.4

Nu   Nusselt number, for convective heat transfer $[-]$; see equation (9.11) and Appendix T

P pressure $[Pa = N/m^2]$; see Appendix G

$P = \Delta T_i/\Delta T_{max}$ temperature change of phase i compared to the maximum possible $[-]$; see Fig. 13.4

$PE = zg/g_c$ shorthand notation for the potential energy of the flowing fluid $[J/kg]$

Pr   Prandtl number for fluids $[-]$; see Appendix T

q heat added to a flowing fluid $[J/kg]$

$\dot{q}$ heat transfer rate $[W]$

$\dot{q}_{12}$ flow rate of energy from surface 1 to surface 2 $[W]$; see equation (9.65)

Q heat lost or gained by a fluid up to a given point in the exchanger $[J/kg$ of a particular flowing phase$]$; see Fig. 13.4

$R = 8.314\ J/mol\,K$ gas constant for ideal gases; see Appendix K

R ratio of temperature changes of the two fluids in an exchanger $[-]$; see Fig. 13.4

Re	Reynolds number for flowing fluids [$-$]; see text after equation (2.4) and Appendix T
$Re = du\rho/\mu$	for flow of Newtonians in pipes; see equation (2.4)
$Re = du\rho/\eta$	for flow of Bingham plastics in pipes; see equation (5.8)
$Re_{gen} = \left(\dfrac{d^n u^{2-n}\rho}{8^{n-1}K}\right)\left(\dfrac{4n}{1+3n}\right)^n$	for flow of power law fluids in circular pipes; see equation (5.10)
$Re_p = d_p u_0 \rho/\mu$	for flow in packed and fluidized beds; see equation (6.9)
$Re_t = d_p u_t \rho/\mu$	at the terminal velocity of a falling particle; see equation (8.6)
S	entropy of an element of flowing fluid [J/kg K]
s	pumping speed, volumetric flow rate of gas at a given location in a pipe [m³/s]; see equation (4.4)
t	time [s]
T	temperature [K]
$\overline{\Delta T}$	proper mean temperature difference between the two fluids in an exchanger [K]
u	velocity or mean velocity [m/s]
u_{mf}	minimum fluidizing velocity [m/s]
u_0	superficial velocity for a packed or fluidized bed, thus the velocity of fluid if the bed contained no solids [m/s]; see equation (6.9)
U	internal energy [J/kg]; see equation (1.1)
U	overall heat transfer coefficient [W/m² K]; see text after equation (10.4)
u_t	terminal velocity of a particle in a fluid [m/s]
$\dot{v}$	volumetric flow rate of fluid [m³/s]
V	volume [m³]
W	mass of a batch of material [kg]
W_{flow}	work done by fluid in pushing back the atmosphere; this work is not recoverable as useful work [J/kg]
W_s	shaft work; this is the mechanical work produced by the fluid which is transmitted to the surroundings [J/kg]
$\dot{W}_s$	pumping power; that produced by fluid and transmitted to the surroundings [W]
y	distance from the wall of a flow channel [m]; first sketch in Chapter 2
z	height above some arbitrarily selected level [m]
Z	compressibility factor, correction factor to the ideal gas law [$-$]; see text after equation (3.15)

Greek Symbols

α	kinetic energy correction factor [$-$]; see equation (2.12)
$\alpha = k/\rho C_p$	thermal diffusivity [m²/s]; see equation (11.1) and Appendices Q and U

α	the absorptivity or the fraction of incident radiation absorbed by a surface [$-$]; see equation (9.59)
ϵ	pipe roughness [m]; see Table 2.1
ε	voidage in packed and fluidized beds [$-$]; see Figs. 6.3 and 6.4
ε	emissivity of a surface [$-$]; see equation (9.60)
η	pump efficiency [$-$]; see equation (1.14)
η	plastic viscosity of Bingham plastic non-Newtonian fluids [kg/m s]; see equation (5.2)
$\eta_i = \Delta T_i / \Delta T_{max}$	efficiency or effectiveness of heat utilization of stream i, or fractional temperature change of stream i [$-$]; see equation (17.3) or (18.2)
π	atmospheric pressure
μ	viscosity of a Newtonian fluid [kg/m s]; see equation (5.1) and Appendix M
ρ	density [kg/m^3]; see Appendices L and U
σ	standard deviation in the spread of the temperature front in a packed bed regenerator; see equation (15.9)
$\sigma = 5.73 \times 10^{-8}$ W/m^2 K^4	Stefan–Boltzmann radiation constant; see equation (12.5)
τ	shear stress (Pa = N/m^2); see text above equation (2.1) and beginning of Chapter 5
τ_w	shear stress at the wall [Pa]
τ_0	yield stress of Bingham plastics [Pa]; see equation (5.2)
ϕ	sphericity of particles [$-$]; see equation (6.1)
$\phi = \dot{m}_g C_g / \dot{m}_s C_s$	heat flow ratio of two contacting streams [$-$]; see equation (14.3)
ϕ'	heat flow ratio for each stage of a multistage contacting unit [$-$]; see equation (14.10)

Subscripts

f	property of the fluid at the film temperature, considered to be the average between the bulk and wall temperatures, or at $T_f = (T_{wall} + T_{bulk})/2$; see equation (9.24)
g	gas
l	liquid
lm	logarithmic mean
s	solid

Part I   FLOW OF FLUIDS
AND MIXTURES

Although the first part of this little volume deals primarily with the flow of fluids and mixtures through pipes, it also considers the flow of fluids through packed beds and through swarms of suspended solids called fluidized solids, as well as the flow of single particles through fluids. The term "fluids and mixtures" includes all sorts of materials under a wide range of conditions, such as Newtonians (e.g., air, water, whiskey), non-Newtonians (e.g., peanut butter, toothpaste), gases approaching the speed of sound, and gas flow under high vacuum where collisions between molecules are rare.

Chapter 1 presents the two basic equations which are the starting point for all analyses of fluid flow, the total energy balance and the mechanical energy balance. Chapters 2 through 8 then take up the different kinds of flow.

Chapter 1 BASIC EQUATIONS FOR FLOWING STREAMS

I. TOTAL ENERGY BALANCE

Consider the energy interactions as a stream of material passes in steady flow between points 1 and 2 of a piping system, as shown in Fig. 1.1. From the first law of thermodynamics we have for each unit mass of flowing fluid:

$$\underbrace{\Delta U}_{\substack{\text{Internal} \\ \text{energy}}} + \underbrace{\Delta\left(\frac{gz}{g_c}\right)}_{\substack{\text{Potential} \\ \text{energy, PE}}} + \underbrace{\Delta\left(\frac{u^2}{2g_c}\right)}_{\substack{\text{Kinetic} \\ \text{energy, KE}}} + \underbrace{\Delta\left(\frac{p}{\rho}\right)}_{\substack{\text{Flow} \\ \text{work}}} = \underbrace{q}_{\substack{\text{Heat added to fluid} \\ \text{from surroundings}}} - \underbrace{W_s}_{\substack{\text{Work received by} \\ \text{surroundings from fluid}}} \quad \left[\frac{\text{J}}{\text{kg}}\right] \qquad (1.1)$$

Consider the internal energy term in the above expression. From the second

3

Fig. 1.1. Energy aspects of a single-stream piping system.

law of thermodynamics

$$\Delta U = \int T\, dS - \int p\, d\left(\frac{1}{\rho}\right) + \begin{array}{l} \text{magnetic,} \\ \text{electrical,} \\ \text{chemical work,} \\ \text{etc.} \end{array} \tag{1.2}$$

The $\int T\, dS$ term accounts for both heat and frictional effects. Thus, in the ideal situation where there is no degradation of mechanical energy (no frictional loss, turbulence, etc.):

$$\int T\, dS = q = \begin{array}{l} \text{heat interchange between} \\ \text{fluid and surroundings} \end{array}$$

On the other hand, in situations where there is degradation (frictional losses)

$$\int T dS = q + \Sigma F \tag{1.3}$$

Heat from surroundings

Total heat added to fluid

Heat generated within the fluid by friction

Noting that $\Delta H = \Delta U + \Delta(p/\rho)$, we can rewrite equation (1.1) as

$$\Delta H + \Delta\left(\frac{gz}{g_c}\right) + \Delta\left(\frac{u^2}{2g_c}\right) = q - W_s \qquad \left[\frac{\text{J}}{\text{kg}}\right] \tag{1.4}$$

[AUTHOR'S NOTE: g_c is a conversion factor, to be used with American engineering units. In SI units, g_c is unity and drops from all equations. Although this book uses SI units throughout, g_c is included in the text and problems for the benefit of those who use American engineering units.]

This is the first law of thermodynamics in its usual and useful form for steady-flow single-stream systems.

II. MECHANICAL ENERGY BALANCE

For each kilogram of real flowing fluid, with its unavoidable frictional effects, with no unusual work effects (magnetic, electrical, surface, or chemical), and with constant value of g, equations (1.1) and (1.3) combined give the so-called "mechanical energy balance"

$$\frac{g}{g_c}\Delta z + \Delta\left(\frac{u^2}{2g_c}\right) + \int \frac{dp}{\rho} + W_s + \Sigma F = 0 \qquad \left[\frac{J}{kg}\right] \qquad (1.5)$$

Along flow path

Work received by surroundings

For incompressible fluids this term becomes $(\Delta p/\rho)$

Friction loss, or mechanical energy transformed into internal energy, and used to heat or vaporize the fluid.

Multiplying by g_c/g gives, in alternative form,

$$\Delta z + \Delta\left(\frac{u^2}{2g}\right) + \frac{g_c}{g}\int \frac{dp}{\rho} + \frac{g_c}{g}W_s + \frac{g_c}{g}\Sigma F = 0 \qquad \left[\begin{array}{c} m\ of \\ fluid \end{array}\right] \qquad (1.6)$$

Lost head $= h_L$

In differential form these equations are

$$\frac{g}{g_c}dz + \frac{u\,du}{g_c} + \frac{dp}{\rho} + dW_s + d(\Sigma F) = 0 \qquad \left[\frac{J}{kg}\right] \qquad (1.7)$$

and

$$dz + \frac{u\,du}{g} + \frac{g_c}{g}\frac{dp}{\rho} + \frac{g_c}{g}dW_s + d(h_L) = 0 \qquad [m] \qquad (1.8)$$

These equations, in fact, represent not a balance, but a loss of mechanical energy (the transformation into internal energy because of friction) as the fluid flows down the piping system. In the special case where the fluid does no work on the surroundings and where the frictional effects are so minor that they can be completely ignored, the mechanical energy balance reduces to

$$\frac{g}{g_c}\Delta z + \Delta\frac{u^2}{2g_c} + \int\frac{dp}{\rho} = 0$$

which is called the Bernoulli equation.

The mechanical energy balance, equations (1.5)–(1.8), is the starting point for finding work effects in flowing fluids—pressure drop, pumping power, limiting velocities, and so on. We apply these expressions to all types of fluids and mixtures.

III. PUMPING ENERGY AND PUMPING POWER

Pumps, compressors, and blowers are the means for making fluids flow in pipes. The shaft work needed is found by writing the mechanical energy

balance about the device. Since kinetic and potential energy losses and piping losses are negligible, equation (1.5) reduces to

$$-W_s = \int_{p_1}^{p_2}\frac{dp}{\rho} \qquad \left[\frac{J}{kg}\right]$$

For *liquids* and *slurries*, $\rho \cong$ const, so the shaft work required is given as

$$-W_s = \frac{p_2 - p_1}{\rho} = \frac{\Delta p}{\rho} \qquad \left[\frac{J}{kg}\right] \tag{1.9}$$

For *gases*, assuming ideal gas behavior and adiabatic reversible compression, the work done on the gas becomes

$$-W_{s,\text{ideal}} = \frac{k}{k-1}\frac{RT_1}{(mw)}\left[\left(\frac{p_2}{p_1}\right)^{(k-1)/k} - 1\right] \tag{1.10}$$

$$= \frac{k}{k-1}\frac{RT_2}{(mw)}\left[1 - \left(\frac{p_1}{p_2}\right)^{(k-1)/k}\right] \quad \left[\frac{J}{kg}\right] \tag{1.11}$$

The power requirement is then

$$-\dot{W}_{s,\text{ideal}} = \frac{k}{k-1}\dot{n}RT_1\left[\left(\frac{p_2}{p_1}\right)^{(k-1)/k} - 1\right] \tag{1.12}$$

$$= \frac{k}{k-1}\dot{n}RT_2\left[1 - \left(\frac{p_1}{p_2}\right)^{(k-1)/k}\right] \quad [\text{W}] \tag{1.13}$$

where

$$k = \frac{C_p}{C_v}, \quad \frac{RT_i}{(mw)} = \frac{p_i}{\rho_i}, \quad \text{and} \quad \dot{n}RT_i = p_i\dot{v}_i$$

Next compare ideal compression with real compression with its frictional losses, both designed to take fluid from a low pressure p_1 to a higher pressure p_2. The p-T diagram of Fig. 1.2 shows the path taken by the fluid in these two cases.

Note that in the real compression the unavoidable frictional effects cause the leaving fluid (point C) to be hotter than it would be ideally (point B). Thus, some of the incoming shaft work is not used for compression, but is used to heat the fluid. Thus the actual shaft work is always greater than the ideal and is related to it by

$$W_{s,\text{actual}} = \frac{W_{s,\text{ideal}}}{\eta} \quad W_s < 0 \tag{1.14}$$

Fig. 1.2. In a real compression the fluid leaves hotter than in an ideal compression.

where p_1 and p_2 used in $W_{s,\text{ideal}}$ are the measured pressures, and η is the compressor efficiency, roughly given by

$$\eta = 0.55 \sim 0.75 \qquad \text{for a turbo-blower}$$

$$= 0.60 \sim 0.80 \qquad \text{for a Roots blower}$$

$$= 0.80 \sim 0.90 \qquad \text{for an axial blower, or a two-stage reciprocating compressor}$$

Consider the reverse operation where the flowing fluid generates work while going from a high to a low pressure, an example being the flow of water through a turbine to generate electricity. In the real turbine with its frictional losses, less work is generated than could have been generated ideally. These two work terms are related by

$$W_{s,\text{actual}} = \eta W_{s,\text{ideal}} \qquad W_s > 0 \tag{1.15}$$

where $W_{s,\text{ideal}}$ is calculated by one of equations (1.9)–(1.13) using the actual high and low pressures.

As we see, in either of these changes a portion of the mechanical energy is lost to friction. This is the unavoidable tax imposed by the second law of thermodynamics.

EXAMPLE 1.1. HYDROSTATICS AND MANOMETERS

Find the pressure p_4 in the tank from the manometer reading shown below, knowing all heights z_1, z_2, z_3, z_4, all densities ρ_A, ρ_B, ρ_C, and the surrounding pressure p_1.

Solution

To find the pressure at point 4, apply the mechanical energy balance from a point of known pressure, point 1, around the system to point 4. Thus, from point 1 to point 2 we have

$$\frac{g\Delta z}{g_c} + \cancelto{=0}{\frac{\Delta u^2}{2g_c}} + \int \frac{dp}{\rho} + \cancelto{=0}{W_s} + \cancelto{=0}{\sum F} = 0$$

and for fluids of constant density this reduces to

$$p_2 - p_1 = \frac{\rho_C g}{g_c}(z_1 - z_2)$$

Repeating this procedure gives

$$p_3 - p_2 = \frac{\rho_B g}{g_c}(z_2 - z_3)$$

$$p_4 - p_3 = \frac{\rho_A g}{g_c}(z_3 - z_4)$$

Adding the above expression and noting that $p_1 = 100$ kPa gives the desired expression

$$p_4 = 100 + \frac{g}{g_c}[\rho_C(z_1 - z_2) + \rho_B(z_2 - z_3) + \rho_A(z_4 - z_3)]$$

The same strategy of working around the system holds for other geometries and for piping loops.

EXAMPLE 1.2. COUNTING CANARIES ITALIAN STYLE

Italians love birds, many homes have these happy songsters in little cages, and to supply them is a big business. Tunisian Songbirds, Inc., is a major supplier of canaries for Southern Italy, and every Wednesday a large truck carrying these chirpy feathered creatures is loaded aboard the midweek Tunis to Naples ferry. The truck's bird container is 2.4 m wide, 3.0 m high, solid on the sides and bottom, open at the top except for a restraining screen, and has a total open volume available for birds of 36 m³. On arrival at Naples a tax of 20 lira/bird is to be charged by the customs agent, but how to determine the amount to be assessed? Since counting these thousands of birds one by one would be impractical the Italians use the following ingenious method.

The customs agent sets up his pressure gauges, then loudly bangs the side of the van with a hammer. This scares the birds off their perches up into the air. Then he carefully records the pressure both at the bottom and at the top of the inside of the van.

If the pressure at the bottom is 103,316 Pa, the pressure at the top is 102,875 Pa, and the temperature is 25°C, how much tax should the customs agent levy?

Additional data: Juvenile canaries have a mass of 15 gm and a density estimated to be 500 kg/m^3.

Solution

With birds flying in all directions, consider the interior of the van to be a "fluid" or "slurry" or a "suspension" of mean density $\bar{\rho}$; then the mechanical energy balance of equation (1.5) reduces to

$$p_1 - p_2 = \frac{\bar{\rho}g(z_2 - z_1)}{g_c} \tag{i}$$

where

$$\bar{\rho} = \frac{\text{total mass}}{\text{total volume}} = \frac{m_{\text{birds}} + m_{\text{air}}}{V_{\text{total}}} = \frac{\rho_b V_b + \rho_a V_a}{V_t}$$

$$= \rho_b \left(\frac{V_b}{V_t}\right) + \rho_a \underbrace{\left(\frac{V_a}{V_t}\right)}_{= 0}$$

In their calculations the customs agents ignore the second term in the above expression because the density of air is so small compared to the density of the birds. Thus, on replacing all known values into equation (i) gives

$$103{,}316 - 102{,}875 = [(500)(V_b/V_t)](9.8)(3 - 0)$$

from which the volume fraction of birds in the van is found to be

$$\frac{V_b}{V_t} = 0.03$$

The number of birds being transported is then

$$N = \left(0.03\,\frac{m^3\ \text{birds}}{m^3\ \text{van}}\right)(36\ m^3\ \text{van})\left(500\,\frac{kg}{m^3\ \text{bird}}\right)\left(\frac{1\ \text{bird}}{0.015\ \text{kg}}\right) = 36{,}000\ \text{birds}$$

therefore

$$\text{Tax} = \left(20\,\frac{L}{\text{bird}}\right)(36000\ \text{birds}) = \boxed{720{,}000\ L}$$

PROBLEMS ON ENERGY BALANCES

1.1. Thermodynamics states that at high enough pressure diamond is the stable form for carbon. General Electric Co. and others have used this information to make diamonds commercially by implosion and various other high-pressure techniques, all complex and requiring sophisticated technology.

Let us try something different. Take a canvas sack full of lead pencils, charcoal briquets, and coal out in a rowboat to one of the deepest parts of the ocean, the Puerto Rico trench; put some iron pipe in the sack, lower it to the bottom 10 km below, wait a day, and then haul it up. Lo and behold!—a sack full of diamonds, we hope. It may work if the pressure on the ocean bottom is above the critical, or transition, pressure. Find the pressure on the ocean bottom.

Data: Down to 10 km, sea water has an average density of 1036 kg/m³.

1.2. (a) For the manometer shown below develop the expression for the pressure difference $p_1 - p_4$ as a function of the pertinent variables.

(b) If fluid A is liquid and B is gas, then $\rho_B \ll \rho_A$ and ρ_B can be ignored. In this special situation show how the general expression simplifies.

1.3. Find the pressure difference $p_1 - p_2$ in terms of Δz for the two-fluid manometer shown below.

1.4. Consider a very high column of gas, isothermal, and with an average compressibility factor $\overline{Z}$.

(a) Show that the pressure difference between the top (point 2) and bottom (point 1) of this column is given by

$$\Delta p = p_1 \left[e^{[(mw)g/\overline{Z}RTg_c](z_1 - z_2)} - 1 \right]$$

(b) A capped natural gas well has a pressure of 15 atm at ground level. Assuming this natural gas to be pure methane, calculate the pressure of gas 5000 m underground if the temperature is 300K, and $\overline{Z} = 0.95$ throughout.

1.5. The two-fluid manometer is used when the flowing fluid A can chemically attack the manometer fluid C and must be kept from it. For the two-fluid manometer shown below, develop an expression for $p_1 - p_6$ in terms of the pertinent variables.

1.6. *Pitot tubes* are simple reliable devices for measuring the velocity of flowing fluids. They are used in the laboratory, and, if you look carefully, you will see them on all airplanes. The diagram below shows how the pitot tube works. The pressure read by probe A is higher than that read by probe B. Fluid flows past probe B but is brought to a stop at probe A, and according to Bernoulli's equation the difference in velocity is translated into a difference in pressure. Real pitot tubes compactly combine these two probes into two concentric tubes as shown here.

(a)   (b)

(a) Develop the general expression for flow velocity in terms of the pressures p_A and p_B.
(b) Titan, Saturn's moon, is the largest satellite in our solar system. It is roughly half the diameter of the earth, its atmosphere consists mainly of methane, and it is probably the easiest object to explore in the outer solar system. As the Voyager 2 spacecraft slowly settles towards the surface of Titan through an atmosphere at $-130°C$ and 8.4 kPa, its pitot tube reads a pressure difference of 140 Pa. Find the speed of the spacecraft.

1.7. A *venturi meter* is a device for measuring the flow rate of fluid in a pipe. It consists of a smooth contraction and expansion of the flow channel, as shown below ($p_1 \gg p_2$). Pressure measurements at the throat and upstream then give the flow rate of fluid. For liquid flowing through an ideal venturi, show that the approach velocity is given by the following equation:

$$u_1 = \frac{1}{(\alpha^4 - 1)^{1/2}} \cdot \left(\frac{2g_c(p_1 - p_2)}{\rho} \right)^{1/2} \quad \text{where} \quad \alpha = \frac{d_1}{d_2}$$

NOTE: For a well-designed venturi, where $d_2 < d_1/4$, this expression is off by 1–2% at most, because of frictional effects. Thus, in the real venturi meter, the "1" in the above expression should be replaced by 0.98 ~ 0.99.

1.8. A *hydraulic ram* uses the kinetic energy of flowing fluid to raise part of that fluid to a higher derivation. The principle of operation is as follows:

When water flowing down a supply tube is abruptly stopped the pressure at the bottom of the pipe surges, and this allows some of the water to be raised above the feed level. With proper valving this pulsing action repeats 15–200 times/min, with pumping efficiencies as high as 90%. This type of pump requires no motor and is convenient to use in far-away places.

(a) Develop an expression relating the fraction of feed which can be delivered to the higher elevation to the pertinent variables shown above.

(b) A stream flows past my home and I want to use it for my domestic water supply. I do not want to buy electricity to pump the water up to my house so I've decided to use a hydraulic ram. If the fall of the stream is 3 m, the usable flow rate in the feed pipe is 2 lit/s, the height from the ram to my storage tank is 8 m, what delivery rate can I expect at 50% efficiency if the efficiency is defined as actual delivery rate divided by the theoretical?

1.9. The *air lift pump* works by pumping small air bubbles into the bottom of a vertical pipe immersed in the fluid to be raised, as shown below. Ideally, the bubbles are so small that the relative velocity between air and water is negligible. Actual air lift pumps can achieve pumping efficiencies of 75% or more. For a 60% water and 40% air mixture in the lift pipe, how high can an ideal air lift pump raise the water?

1.10. A pilot-scale fluidized bed combustor needs a continuous supply of 4 m³/s of 120 kPa air. The room temperature is 293 K.
(a) What size of compressor operating at 74% efficiency will do the job?
(b) What will be the temperature of the air leaving the compressor if no heat is lost to the surroundings?

1.11. I want to get into the *Guinness Book of Records* by constructing the world's highest fountain. For this I will pump water out of a 2-cm-i.d. pipe pointing straight up. Find the compressor power needed for the stream of water to rise

100 m. Now people tell me that I have to consider frictional effects, but I've got an ace up my sleeve. A most honest-looking fellow at the race track sold me a very, very special powder. One pinch into the water supply tank and friction in the pipe and in the air magically disappears. So please don't worry about friction, just make your calculations assuming that frictional effects can be ignored.

1.12. (a) Referring to Example 2 of this chapter, the importers suspect that the simplifying assumption being used by Italian customs is unfair to them. To check this, determine the tax to be paid if the air density term was accounted for in the calculations.

   (b) If the pressure reading of 102,875 Pa were actually taken outside the van and at the same level as the 103,316 Pa reading, what would be the proper tax to be paid?

Chapter 2   FLOW OF INCOMPRESSIBLE NEWTONIAN FLUIDS IN PIPES

Newtonians are fluids in which the relative slip of fluid elements past each other is proportional to the shear on the fluid, as shown in Fig. 2.1. All gases, liquid water, and liquids of simple molecules (ammonia, alcohol, benzene, oil, chloroform, butane, etc.) are Newtonians. Pastes, emulsions, biological fluids, polymers, suspensions of solids, and other mixtures are likely to be non-Newtonian.* This chapter deals with Newtonians.

When a fluid flows in a pipe some of its mechanical energy is dissipated by friction. The ratio of this frictional loss to the kinetic energy of the flowing fluid is defined as the Fanning friction factor, f_F. Thus[†]

$$ f_F = \left(\frac{\text{frictional drag force} \Big/ \text{area of pipe surface}}{\text{kinetic energy}/\text{m}^3 \text{ of fluid}} \right) = \frac{\tau_w}{\rho \dfrac{u^2}{2g_c}} \qquad [-] \qquad (2.1) $$

*Strictly speaking, we should use the terms *Newtonian fluids* and *non-Newtonian fluids*. However, there should not be much confusion or distress if we drop the word "fluid" and simply call these materials *Newtonians* and *non-Newtonians*.
[†]SI readers may ignore g_c in all equations, if they wish.

Fig. 2.1. Representation of a Newtonian fluid.

Making a force balance about a section of pipe, as sketched in Fig. 2.2, relates the wall shear τ_w to the frictional loss ΣF (or the frictional pressure drop Δp_{fr}). In words then

$$\begin{pmatrix} \text{Force transmitted} \\ \text{to the walls} \end{pmatrix} = \begin{pmatrix} \text{frictional energy} \\ \text{loss by the fluid} \end{pmatrix}$$

or in symbols

$$(L\pi d)\,\tau_w = \frac{\pi d^2}{4}\Delta p_{\text{fr}} = \frac{\pi d^2}{4}\rho\Sigma F \qquad [\text{N}]$$

Substituting in equation (2.1) and rearranging gives

$$\Sigma F = \frac{2f_F L u^2}{g_c d} = \frac{g}{g_c}h_L \qquad \left[\frac{\text{J}}{\text{kg}}\right] \tag{2.2}$$

This frictional loss shows up as generated heat, or $\Sigma F = Q_{\text{generated}}$, and the lost head is seen physically in the sketch of Fig. 2.3.

We would expect the friction factor to depend on the velocity of flow u, the fluid properties of density ρ and viscosity μ, the pipe size d, and its roughness, and so it does. Thus, we have

$$f_F = f\left[\begin{pmatrix} \text{Reynolds number:} \\ \text{a combination of } d, u, \mu, \rho \end{pmatrix}, \begin{pmatrix} \text{pipe} \\ \text{roughness, } \epsilon \end{pmatrix}\right]$$

Fig. 2.2. A force balance on a section of pipe.

Fig. 2.3. Physical representation of the lost head.

where the Reynolds number is defined as

$$\mathrm{Re} = \frac{du\,\rho}{\mu}$$

Figures 2.4 and 2.5, prepared from both theory and experiment, represent this relationship for Newtonians in two alternative ways. Each figure is useful for certain purposes. The pipe roughness, needed in these charts, is given in Table 2.1 for various common pipe materials.

The mechanical energy balance for flow between points 1 and 2 in a pipe is then represented by equation (1.5); thus, referring to Fig. 2.6 we have

where

$$\left.\begin{array}{c} \dfrac{g}{g_c}\Delta z + \Delta\left(\dfrac{u^2}{2g_c}\right) + \displaystyle\int_1^2 \dfrac{dp}{\rho} + \overbrace{W_s}^{\substack{=\,0\text{ for no pump or turbine}\\ \text{in the line}}} + \Sigma F = 0 \\[2mm] \Sigma F = \dfrac{2f_F Lu^2}{g_c d} \end{array}\right\} \quad \left[\dfrac{\mathrm{J}}{\mathrm{kg}}\right] \qquad (2.3)$$

or in terms of head of fluid

where

$$\left.\begin{array}{c} \Delta z + \Delta\left(\dfrac{u^2}{2g}\right) + \dfrac{g_c}{g}\displaystyle\int \dfrac{dp}{\rho} + \dfrac{g_c}{g}\overbrace{W_s}^{\substack{=\,0\text{ for no pump or turbine}\\ \text{in the line}}} + h_L = 0 \\[2mm] h_L = \dfrac{g_c}{g}\Sigma F = \dfrac{2f_F Lu^2}{gd} \end{array}\right\} \quad [\mathrm{m}] \qquad (2.4)$$

These equations are used with Figs. 2.4 and 2.5 to solve pipe flow problems.

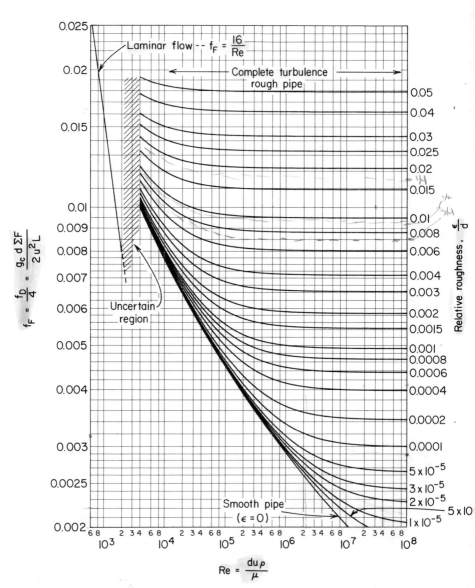

Fig. 2.4. This figure is useful for finding the pumping requirement or frictional loss when you are given the flow rate of fluid in a pipe. [Adapted from Moody (1944).]

Fig. 2.5. This figure is useful for finding the flow rate when you are given the driving force for flow (gravitational head, pumping energy input, etc.). [Adapted from H. Rouse; see discussion after Moody (1944).]

Table 2.1. Roughness of Clean Pipe[a]

Pipe material	ϵ, mm
Riveted steel	1–10
Concrete	0.3–3
Wood stave	0.2–1
Cast iron	$0.26\ (0.25)^b$
Galvanized iron	$0.15\ (0.15)^b$
Asphalted cast iron	$0.12\ (0.13)^b$
Commercial steel or wrought iron	$0.046\ (0.043)^b$
Drawn tubing	0.0015
Glass	0
Plastic (PVC, ABS, polyethylene)	0

[a]Adapted from Fischer and Porter Co., Hatboro, PA, Catalogs section 98-A (1947).
[b]Values in parentheses from Colebrook (1939).

A. Comments

1. **The Reynolds number Re** measures the importance of energy dissipation by viscous effects. Thus

$$\text{Re} = \frac{\text{inertial forces}}{\text{viscous forces}} = \frac{du\,\rho}{\mu} = \frac{dG}{\mu}$$

When the flow is represented by a large Reynolds number this means that viscous effects are relatively unimportant and contribute little to energy dissipation; a small Reynolds number means that viscous forces dominate and are the main mechanism for energy dissipation.

2. **Flow regimes (see Fig. 2.7).** Newtonians flowing in pipes exhibit two distinct types of flow, laminar (or streamline) when Re < 2100, and turbulent when Re > 4000. Between Re = 2100 and Re = 4000 we observe a transition regime with uncertain and sometimes fluctuating flow.

Fig. 2.6. Development of the mechanical energy balance for flow in pipes.

Fig. 2.7. The difference between laminar and turbulent flow in pipes.

3. **For laminar flow** (Re < 2100), the friction factor and the frictional loss can be found either from Figs. 2.4 or 2.5, or from the following simple theoretical expressions derived by Poiseuille

$$f_F = \frac{16}{Re} \quad \text{or} \quad \Sigma F = \frac{32u\mu L}{g_c d^2 \rho} \quad \text{or} \quad u = \frac{g_c d^2 \rho \, \Sigma F}{32\mu L} \qquad (2.5)$$

4. **Two different friction factors** are in common use today:

(i) f_F, the Fanning friction factor, defined in equation (2.1);
(ii) f_D, the Darcy friction factor.

Chemical engineers favor f_F; most others prefer f_D. Don't confuse these two definitions; also, note that

$$f_D = 4f_F \quad \text{or} \quad f_F = \frac{f_D}{4}$$

The simplest way to tell which one is being used (when no subscript is shown) is to look at the laminar flow regime in the f vs. Re chart. There

$$f_F = \frac{16}{Re} \quad \text{while} \quad f_D = \frac{64}{Re}$$

5. **In the turbulent regime** (Re > 4000), the friction factor and frictional loss are found from Figs. 2.4 or 2.5, or from the experimentally tested expressions of Nikuradse, which closely approximate the theoretical expressions of Nikuradse, Prandtl, and von Karman, as discussed in Schlichting (1979).

For the range of flows from Re = 4000 to 10^8 these expressions were cleverly combined by Colebrook (1939) to give

$$\frac{1}{\sqrt{f_F}} = -4\log\left(\frac{1}{3.7}\frac{\epsilon}{d} + \frac{1.255}{Re\sqrt{f_F}} \right) \qquad (2.6)$$

or in a form useful for calculating Re given the value of f_F

$$\frac{1}{\text{Re}} = \frac{\sqrt{f_F}}{1.255}\left[10^{-0.25/\sqrt{f}} - \frac{1}{3.7}\frac{\varepsilon}{d}\right] \tag{2.7}$$

Note that in these expressions the friction factor can only be calculated by trial and error. But with our handy helper, the hand calculator, this should not be too tedious; just guess a few f_F values and interpolate.

The Colebrook equation reduces to a number of special cases. Thus for *fully developed turbulent flow in rough pipes*, where f_F is independent of Re, it simplifies to

$$\frac{1}{\sqrt{f_F}} = 4\log\left(3.7\frac{d}{\varepsilon}\right) \tag{2.8}$$

For *smooth pipes* ($\varepsilon/d = 0$) it simplifies to

$$\frac{1}{\sqrt{f_F}} = 4\log\frac{\text{Re}\sqrt{f_F}}{1.255} \tag{2.9}$$

Alternatively, for smooth pipes and Re < 100,000, we have the simpler Blasius expression as an approximation to equation (2.9)

$$f_F = 0.079\,\text{Re}^{-1/4} \tag{2.10}$$

6. **Transition regime** (Re = 2100 ~ 4000). Here we have an uncertain situation where the flow may be turbulent, or laminar, or even fluctuating.

7. **Piping systems** have contractions, expansions, valves, elbows, and all sorts of fittings. Each has its own particular frictional loss. A convenient way to account for this is to put this loss in terms of an equivalent length of straight pipe. Thus, the equivalent length of a piping system as a whole is given by

$$L_{\substack{\text{equiv}\\ \text{total}}} = L_{\substack{\text{straight}\\ \text{pipe}}} + \Sigma\left(L_{\substack{\text{equiv}\\ \text{all fittings,}\\ \text{contractions,}\\ \text{expansions,etc.}}}\right) \tag{2.11}$$

In turbulent flow the equivalent lengths of pipe fittings are independent of the Reynolds number, and Table 2.2 shows these values for various fittings. Unfortunately, in laminar flow the equivalent length varies strongly with the

Table 2.2. Equivalent Pipe Length for Various Pipe Fittings
(Turbulent Flow Only)[a]

Pipe fitting	L_{equiv}/d
Globe valve, wide open	~ 300
Angle valve, wide open	~ 170
Gate valve, wide open	~ 7
3/4 open	~ 40
1/2 open	~ 200
1/4 open	~ 900
90° elbow, standard	30
long radius	20
45° elbow, standard	15
Tee, used as elbow, entering the stem	90
Tee, used as elbow, entering one of two side arms	60
Tee, straight through	20
180° close return bend	75
Ordinary entrance (pipe flush with wall of vessel)	16
Borda entrance (pipe protruding into vessel)	30
Rounded entrance, union, coupling	Negligible
Sudden enlargement from d to D	
Laminar flow in d:	$\dfrac{Re}{32}\left[1-\left(\dfrac{d^2}{D^2}\right)\right]^2$
Turbulent flow in d:	$\dfrac{1}{4}f_{F,in\,d}\left[1-\left(\dfrac{d^2}{D^2}\right)\right]^2$
Sudden contraction from D to d; all conditions except high-speed gas flow where $p_1/p_2 \geq 2$. For this see Chapter 3.	
Laminar flow in d:	$\dfrac{Re}{160}\left[1.25-\left(\dfrac{d^2}{D^2}\right)\right]$
Turbulent flow in d:	$\dfrac{1}{10}f_{F,in\,d}\left[1.25-\left(\dfrac{d^2}{D^2}\right)\right]$

[a]Adapted in part from Crane (1982) and from Perry (1950).

Reynolds number in distinctive ways for each fitting. Thus, simple generalizations such as Table 2.2 cannot be prepared for the laminar flow regime.

8. The kinetic energy of flowing fluids. In solving flow problems and replacing values in the mechanical energy balance we often find

- for liquids—the kinetic energy terms are negligible and can be ignored;
- for gases—the potential energy terms are negligible and can be ignored.

For a flat
velocity profile $\cdots KE = {}^{u^2}\!/2g_c$
(plug flow)

For laminar flow
(parabolic $\cdots KE = 2({}^{u^2}\!/2g_c) = {}^{u^2}\!/g_c$
velocity profile)

For turbulent flow $\cdots KE = 1.02 \sim 1.15 ({}^{u^2}\!/2g_c)$

u = mean velocity

Fig. 2.8. An accounting of the kinetic energy term in the mechanical energy balance.

When in doubt, evaluate all the terms and then drop those which are small compared to the others.

In cases where the kinetic energy must be considered the sketches of Fig. 2.8 show how to account for this effect. In general, then

$$KE = \frac{u^2}{\alpha 2 g_c} \quad \text{where} \quad \alpha \begin{cases} = \frac{1}{2} \text{ for laminar flow} \\ = 1 \text{ for plug flow} \\ \to 1 \text{ for turbulent flow} \end{cases} \quad (2.12)$$

Since the kinetic energy of flowing fluids only contributes significantly at high velocities where flow is turbulent, α usually is close to 1. Only for gases can the flow be both laminar and at high velocity. This situation occurs only rarely.

9. Evaluation of the $\int (dp/\rho)$ term in the mechanical energy balance:

• for liquids, $\rho \simeq$ constant so

$$\int_1^2 \frac{dp}{\rho} = \frac{1}{\rho}(p_2 - p_1) = \frac{\Delta p}{\rho} \quad (2.13)$$

• for ideal gases with small density changes we can use an average density and then treat the gas as an incompressible fluid; thus

$$\bar{\rho} = \frac{1}{\bar{v}} = \frac{\bar{p}(mw)}{R\bar{T}} \quad \text{where} \quad \begin{cases} \bar{p} = \dfrac{p_1 + p_2}{2} \\ \bar{T} = \dfrac{T_1 + T_2}{2} \end{cases} \quad (2.14)$$

• for large fractional changes in pressure or density, and by this we mean when $p_1/p_2 > 2$ or $\rho_1/\rho_2 > 2$, we must use the treatment of Chapter 3.

10. Aging of pipes. The value of pipe roughness given in Table 2.1 is for clean pipe. With time, however, roughness may increase because of corrosion and scale deposition. Colebrook (1939) found that a simple linear expression can reasonably represent such a change

$$\epsilon_{\text{any time}} = \epsilon_{\text{time}=0} + \alpha t \qquad (2.15)$$

An increase in roughness will lower the flow rate for a given driving force or will increase the power requirement to maintain a given flow rate.

11. Other shaped conduits. In general, one can approximate the frictional loss in other than circular shaped conduits by representing the conduit by a circular pipe of equivalent diameter defined as

$$d_e = 4\left(\frac{\text{hydraulic}}{\text{radius}}\right) = 4\left(\frac{\text{cross-sectional area}}{\text{wetted perimeter}}\right) \qquad (2.16)$$

For certain shapes—

- plain and eccentric annuli;
- pipes containing various shapes of internals, including finned tubes;
- parallel plates; and
- rectangular, triangular, and trapezoidal conduits;

—experiments have been made and frictional losses reported [see Knudsen and Katz (1958), Chapters 4 and 7].

REFERENCES AND RECOMMENDED READINGS[+]

C. F. Colebrook, Friction factors for pipe flow, *Inst. Civil Eng.* **11**, 133 (1939).

Crane Co. Technical Paper 410, New York (1982).

J. G. Knudsen and D. L. Katz, *Fluid Dynamics and Heat Transfer*, McGraw-Hill, New York (1958).

R. Lemlich, A kinetic analogy, *J. Chem. Educ.* **31**, 431 (1954).

L. F. Moody, Turbulent flow in pipes, with particular reference to the transition region between the smooth and the rough pipe law, *Trans. ASME* **66**, 641 (1944).

J. H. Perry, *Chemical Engineer's Handbook*, Third Ed., McGraw-Hill, New York (1950).

H. Schlichting, *Boundary Layer Theory*, Seventh Ed. (transl. by J. Kestin), McGraw-Hill, New York (1979).

G. T. Zhang and S. D. Hau, A grand scale: $12 for a dozen experiments in CRE, *Chem. Eng. Educ.*, p. 10 (Winter 1984).

EXAMPLE 2.1. TOMATO GROWING IN ABSENTIA

Every summer I carefully grow a giant tomato plant because I love the taste of its fresh-picked fruit. Since these plants need 2 liters of water each day of the growing season to produce these delectable and irresistable fruit, how do I grow my plant next summer when I will be away for four weeks with no way to water it?

One solution would be to connect a long plastic tube 0.4-mm-i.d. to the faucet at my home where the water pressure is 100 kPa above atmospheric and lead it to the plant. Determine how long the tube would have to be to deliver 2 lit/day of water. Of course, everything is on the level.

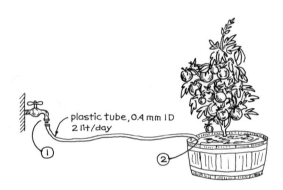

plastic tube, 0.4 mm ID
2 lit/day

Solution

Knowing the volumetric flow rate of water and the tube diameter will allow calculation of the Reynolds number for the flow in the plastic tube. Thus

$$u_2 = \frac{\dot{v}}{A} = \frac{\left(2\dfrac{\text{lit}}{\text{day}}\right)\left(\dfrac{1\ \text{m}^3}{1000\ \text{lit}}\right)\left(\dfrac{1\ \text{day}}{24 \times 3600\ \text{s}}\right)}{\dfrac{\pi}{4}(0.0004\ \text{m})^2} = 0.184\ \frac{\text{m}}{\text{s}}$$

and

$$Re_2 = \frac{d_2 u_2 \rho}{\mu} = \frac{(0.0004)(0.184)(1000)}{10^{-3}} = 73.7 \quad \text{thus, laminar flow}$$

After these preliminaries let us write the mechanical energy balance between points

1 and 2

$$\underbrace{\frac{g(z_2 - z_1)}{g_c}}_{=0} + \frac{u_2^2 - u_1^2}{g_c} + \frac{p_2 - p_1}{\rho} + \underbrace{W_s}_{} + \frac{2f_F L u^2}{g_c d} = 0$$

Note that "2" is absent from the denominator of the kinetic energy term. This is because the fluid is in laminar flow. Also, since the diameter of the faucet opening is large compared to that of the tube, we can reasonably assume that the velocity therein is negligible, or $u_1 = 0$. Thus, on replacing values we find

$$\frac{(0.184)^2}{(1)} + \frac{101,325 - 201,325}{1000} + \frac{2f_F L (0.184)^2}{1(0.0004)} = 0$$

or

$$0.034 - 100 + 169.6 f_F L = 0 \tag{i}$$

Then, either from an extrapolation of Fig. 2.4 or from equation (2.5), we find that

$$f_F = \frac{16}{Re} = \frac{16}{73.7} = 0.2171 \tag{ii}$$

Combining (i) and (ii) then gives the length of tube needed, or

$$L = \boxed{2.715 \text{ m}}$$

NOTE: The numbers in equation (i) show that the kinetic energy contributes less than 0.04% to the total energy loss. Thus, the kinetic energy term could very well have been ignored in this problem. This frequently is the case, especially when flow velocities are not great and when frictional losses are severe.

EXAMPLE 2.2. OVERFLOW PIPE FOR A DAM

Next summer I plan to dam Dope Creek to form a little lake. Building the dam is straightforward; however, I haven't yet figured out what size of galvanized pipe to use for the water overflow. The dry season is no problem, it is the wet season with its thunderstorms and flash floods that worries me. My personal meteorologist estimates that in the very worst conceivable situation flow in Dope Creek can reach $\pi/2$ m^3/s.

If the pipe is too small, the water level will rise too high (more than 1 m above the water intake) and will overflow and damage the dam. How large a pipe is needed to guarantee that the water level does not rise above this danger point? The equivalent length of this overflow pipe is 19.6 m, and its discharge is located 3 m below its intake.

Solution

We can choose to write the mechanical energy balance between points 1 and 3 or between points 2 and 3. The former pair seems simpler because then $\Delta p = 0$. Thus, between points 1 and 3, we have

$$\frac{g\Delta z}{g_c} + \cancelto{0}{\frac{\Delta u^2}{2g_c}} + \cancelto{0}{\frac{\Delta p}{\rho}} + \cancelto{0}{W_s} + \Sigma F = 0$$

In Example 2.1 we found that the kinetic energy contribution was negligible so let us start by assuming that it is negligible here as well (we will check this later); and with no pump or turbine in the line the shaft work term disappears. On replacing values we find that

$$9.8(-4) + 0 + 0 + 0 + \frac{2f_F u_2^2(19.6)}{(1)d} = 0 \tag{i}$$

Next, relate the flow velocity and Reynolds number with the pipe diameter as follows

$$u_2 = \frac{\dot{v}}{A} = \frac{\pi/2}{(\pi/4)d^2} = \frac{2}{d^2} \tag{ii}$$

$$Re = \frac{du_2\rho}{\mu} = \frac{d(2/d^2)(1000)}{10^{-3}} = \frac{2 \times 10^6}{d} \tag{iii}$$

Combining (i) and (ii) gives

$$f_F = d^5/4 \qquad\qquad \text{(iv)}$$

Now as d changes, so does Re and f_F, and these in turn are related by Fig. 2.4. So let us solve for the pipe diameter by trial and error, as shown

Guess d	Re [from (iii)]	ϵ/d [from Table 2.1]	f_F [from Fig. 2.4]	f_F [from (iv)]
0.1	2×10^7	0.015	5.4×10^{-3}	0.0025×10^{-3}
0.4	5×10^6	0.0038	4.0×10^{-3}	2.6×10^{-3}
0.44	4.6×10^6	0.00034	3.9×10^{-3}	$4.1 \times 10^{-3}\ldots$ close enough

Therefore, the pipe diameter needed is $d = 0.44$ m.

Finally, let us check to see if we are justified in assuming that the kinetic energy contribution can be ignored. Inserting equation (ii) into the kinetic energy term of equation (i) gives

$$\frac{\Delta u^2}{2g_c} = \frac{u_2^2}{2g_c} = \frac{2}{d^4} = 53.36 \frac{\text{J}}{\text{kg}}$$

Not only is this significant, it is even larger than the potential energy term in equation (i). So our original guess was incorrect and we must include the kinetic energy term in the mechanical energy balance. With this, the modified equation (i) becomes

$$9.8(-4) + \frac{u_2^2}{2} + \frac{2f_F u_2^2 (19.6)}{(1)d} = 0$$

or

$$-39.2 + \frac{2}{d^4} + 156.8\frac{f_F}{d^5} = 0 \qquad\qquad \text{(v)}$$

Repeating the trial-and-error procedure, but using equation (v) in place of equation (iv), we find the pipe diameter to be

$$d = 0.53 \text{ m}$$

NOTES: Normally, in rather long small-diameter pipes, the frictional resistance to flow is large enough so that the kinetic energy term can be ignored. In this problem we should have suspected right away that this was not the case and we should have immediately included the kinetic energy term. It would have saved time and effort. Next time we will know better.

Figures 2.4 and 2.5 allow us to solve flow problems without trial and error whenever frictional losses, flow rate, or pipe length are the unknowns. Unfortunately, when the pipe diameter is unknown, one needs to resort to trial-and-error procedures, as we have seen in this problem. Laminar flow is the exception, for then no trial and error is needed.

PROBLEMS ON INCOMPRESSIBLE FLOW IN PIPES

2.1. When a fluid flows through a pipe at velocity u, some of its mechanical energy is dissipated into internal energy by frictional effects. Let this be ΣF (J/kg). What happens to this frictional energy loss if the flow rate is tripled?
   (a) Assume $\text{Re}_{\text{initial}} = 100$ in a rough pipe, $\epsilon/d = 0.01$.
   (b) Assume $\text{Re}_{\text{initial}} = 10,000$ in a smooth pipe.
   (c) Assume $\text{Re}_{\text{initial}} = 10^6$ in a rough pipe, $\epsilon/d = 0.01$.

2.2. A 100%-efficient 1-kW pump-motor lifts water at 1.6 lit/s from a lake through 80 m of flexible hose into a tank 32 m up on a hill. A second pump with the same length of hose will be used to pump water from the lake at the same rate into a reservoir at ground level. What size of 100%-efficient pump-motor is needed?

2.3. I have just seeded a new 0.6 acre lawn and the county extension agent recommends that I sprinkle it with 2 gal/ft^2 day of water to make it grow properly. To do this I bought a pump, 80 ft of $\frac{1}{2}$-in.-i.d. plastic hose (smooth inside), and a sprinkler, which then are all connected. The pump draws water from a stream whose level is 10 ft below the pump, the water intake is a large polyethylene pipe (negligible resistance) and the lawn is 30 ft above the pump. When the pump is working the pressure gauge at the pump outlet reads 80 psig, my electricity receipts show that I pay 2.3¢ /kW hr, and I must sprinkle for 32 days to develop a healthy lawn, so says the agent. If the daytime temperature of water (when I plan to sprinkle) is 80°F and the pump is 25% efficient find the cost of electricity for this 32-day job.

   NOTE: This is the only non-SI problem in the book.

2.4. Air at rest, at 20°C and 100 kPa, is forced by means of a fan through a horizontal galvanized flue 1 m in diameter and 10 m long at a velocity of 10 m/s. What size of motor should be used if the motor is 90% efficient and the fan is 20% efficient?

2.5. 3 kg/s of 75% H_2SO_4 ($\rho = 1650$ kg/m^3, $\mu = 8.6 \times 10^{-3}$ kg/m s) are to be pumped from one open tank to another through 600 m (total equivalent length which includes bends, fittings, etc.) of 50-mm-i.d. pipe ($\epsilon = 0.046$ mm). The outlet of the first tank is 7 m below its surface. In the second tank the inlet is 2 m below the surface, which in turn is 15 m above the surface of the first tank. Find the power required for this operation if the pump-motor is 50% efficient.

Handwritten annotations at top:
# of power = work done = work done + work done one power
stations while line stations
stations divided the line

2.6. A steel pipe 0.5-m-i.d. is to carry 18000 m^3/day of oil from an oil field to a refinery located 1000 km away. The difference in elevation of the two ends of the line is negligible.

   (a) Calculate the power required to overcome friction in the pipe line.

   (b) Since the maximum allowable pressure in any section of the line is 4 MPa (about 40 atm), it will be necessary to have pumping stations at suitable intervals along the pipe line. What is the smallest number of pumping stations required?

   Data: At the temperature involved the oil has a viscosity of 0.05 kg/m s and a density of 870 kg/m^3.

2.7. Water at 20°C is to flow out of a settling pond to a drain trench through 100 m equivalent length of 100-mm-i.d. galvanized pipe. The level of the pond is 10 m above the discharge end of the pipe. Find the flow rate of water in m^3/min.

2.8. Water at 20°C flows from the base of a large storage tank through a smooth horizontal pipe (100-mm-i.d., 1 km in length) at a velocity of 1 m/s. That is not fast enough. How much pipe must I saw off to get 2.5 times the velocity through the pipe? Ignore kinetic energy and entrance effects.

2.9. *Taming the Mekong*. The giant Mekong River runs from the Himalayas through Southeast Asia, and the Mekong Development Project proposed that 35 dams be built along the river to tap its vast hydroelectric potential. One of these dams, the Pa Mong, is to stand 100 m high, be 25% efficient in overall conversion to electricity, and have an annual output of 20 billion (U.S.) kW hr of electrical energy. Preliminary design by engineer Kunmith Ping suggests using 25 water intakes, each leading to a turbine located 100 m below the upstream reservoir through 200 m of concrete pipe. The guaranteed total flow rate of water to the turbines would be 14,800 m^3/s. Find the size of pipes needed. [See *National Geographic* **134**, 737 (1968).]

2.10. *OTEC: Ocean Thermal Energy Conversion*. In these times of high energy demand let us consider tapping the energy of the ocean, using deep cold water to condense a working fluid which is vaporized in warm surface water. For a 100-MW power plant we need to pump 2.4×10^5 kg/s of cold seawater at 5°C from a depth of 750 m in a 12-m-i.d. pipe to the surface of the ocean. The vertical temperature profile in a tropical ocean can reasonably be approximated by a step function: 25°C down to 250 m, 5°C below. Find the pumping energy needed for a 40%-efficient pump-motor, and find the fraction of the output from the power plant that this represents.

   Data: For ocean water: $\rho(5°C) = 1027.7$ kg/m^3, $\rho(25°C) = 1023.4$ kg/m^3. Information from *Thermal Energy from the Sea* by A. Hagen, Noyes Data Corporation (1975); courtesy H. A. Mitchell.

2.11. The Alaska Pipeline pumps crude oil from Prudhoe Bay to Valdez in a 1.22-m-i.d. pipe, 1270 km long at pressures as high as 8 MPa. The crude is at 50°C and the flow rate of the line is 2.2 m^3/s. Calculate:

   (a) the theoretical number of pumping stations needed;

   (b) the kW rating of the pump-motors set.

NOTE: This data is from *National Geographic* **150**, 684 (1976). In addition, estimate for crude oil at this temperature: $\rho = 910 \text{ kg/m}^3$, $\mu = 6 \times 10^{-3}$ kg/m s.

Geothermal Energy. Northwest Natural Gas Co. is doing exploratory work for a possible $50,000,000 geothermal development to supply Portland, Oregon with hot water from Mt. Hood. Wells would be drilled on the slopes of Mt. Hood, 760 m above Portland to obtain 74°C hot water at 1 atm. This would then be piped to Portland at 1.6 m³/s in a 1.1-m-i.d. pipe 70 km in length.

2.12. Calculate the required size of motor and the pumping cost assuming 50% efficiency for the pump-motor and 2¢/kW hr for electricity; or else, if no pump is needed, find the water pressure in the pipeline at Portland.

2.13. We would like the water pressure at Portland to be 700 kPa, and this can be obtained without pumps by choosing the proper pipe size. What size would do the job? [See *Corvallis Gazette-Times* (September 28, 1977).]

2.14. Water at 10°C is to flow out of a large tank through a 30-gauge 8-in. commercial steel (0.205 m i.d.) piping system with valve open, as shown below. What length of pipe could be used while still maintaining a flow rate of 0.2 m³/s?

2.15. In the above problem replace the globe valve with a gate valve. What does this do to your answer?

2.16. The village of Schaffzell high in the Swiss Alps operates its own modest hydroelectric plant which produces electricity continuously whether it is used or not. When not needed the 240 kW of electricity runs a motor-turbine at 75% efficiency which pumps 5°C water at 0.1 m³/s through an equivalent length of 780 m of pipe to a little lake located 153 m uphill.

When extra electricity is needed the flow is reversed, water runs downhill at the same flow rate, 0.1 m³/s, from the little lake through the turbine to generate the needed electricity, again at 75% efficiency. How much power can be generated with this downflow from the little lake?

2.17. What size of pipe was used by the village of Schaffzell in their system that pumps water up to the little lake (see previous problem)?

2.18. *Drip irrigation* is a means of getting water directly to growing plants with very little waste. One method uses a large-diameter polyethylene "mother" tube (10–15-mm-i.d.) from which lead many small-diameter polyethylene feeder tubes, called drippers, which go directly to the individual plants. What diameter and length of dripper should we use if we want a flow rate of 4 lit/hr to each plant, and if the desired dripper length is to be somewhere between 0.5 and 1.5 m?

    Data: The temperature is 20°C; mother tube pressure = 200 kPa; ambient pressure = 100 kPa; drippers are manufactured in three diameters: 0.5 mm, 1.0 mm, and 1.5 mm.

2.19. One of the first experiments in the undergraduate chemical reaction engineering laboratory is to follow the drop in concentration of material which is reacting away in a batch reactor, as shown below,

and then to represent this by a rate equation of the form

$$\left(\begin{array}{c} \text{Rate of} \\ \text{reaction} \end{array} \right) = -\frac{dC}{dt} = kC^n \quad \text{where } k \text{ and } n \text{ are constants} \qquad (i)$$

    In principle, this is a simple experiment. In practice, this can be a headache, messing around with reagents, standardizing solutions, titrating, etc. To bypass these problems, Lemlich (1954) and Zhang and Hau (1984) have proposed a hydraulic analog to this experiment. This consists of a burette and a capillary as shown below. All one does is fill the burette with water and then follow the drop of water level with time.

On representing the reactant concentration by the height of water, or $C \Leftrightarrow h$, we then can write

$$\left(\begin{array}{c} \text{Rate of} \\ \text{reaction} \end{array}\right) = -\frac{dh}{dt} = kh^n \qquad \text{(ii)}$$

and for a small-diameter capillary we can assume that laminar flow prevails throughout.

(a) Find the constants k and n in terms of the flow parameters of the system L, d, D.

(b) For water at 20°C, $d = 1$ mm, $D = 1$ cm, and $L = 10$ cm, find what equation (ii) reduces to.

(c) For the system of part (b) how long would it take for the water level in the burette to drop from $h = 100$ cm to $h = 50$ cm?

2.20. Consider Problem 2.4 on the horizontal flue. If the flue were vertical and flow were upward, would you get a different answer? Would you find that the size of motor changes from 3 to 8 kW? If so, then what would happen for downflow in the flue? Would you be generating 2 kW of useful work for free? Try to resolve the dilemma of a vertical flue.

2.21. *The Aquatrain.* Low-sulfur coal, highly prized by electric utilities because it needs no pretreatment for sulfur removal, is mined in Colorado. Los Angeles needs this coal, and the slurry pipeline is one way to transport this coal from mine to city. Some claim that this method of transportation is cheaper than by railroad. However, there is one overwhelming drawback to this method—the use of fresh water:

- Fresh water must be used in the slurry; otherwise salts will remain with the powdered coal to cause corrosion problems when the coal is burned.
- On arrival at Los Angeles the fresh water will be contaminated and not usable further.
- Fresh water is scarce in the Southwest United States and too valuable to be used this way.

W. R. Grace and Co. has an alternative proposal called the Aquatrain: the powdered coal is placed in giant cylindrical plastic bags 5 m long and 0.75 m in diameter and the bags are then pumped along with water in a 0.91-m-i.d. pipeline all the way from their mine near Axial, Colorado (elevation = 2000 m) to Los Angeles (at sea level), 2000 km away. It is reported that 15 million tons of coal can be transported per year this way.

This method has another big advantage. At Glenwood Springs some very salty water enters the Colorado River, significantly raising the saline content of the river. By using this saline water in the pipeline instead of fresh river water one can keep away 250,000 tons/yr of salt from the river, thereby reducing the load on the desalination plants presently being built downstream on the Colorado River.

Suppose the pipeline operates 360 days in the year, and that the plastic baggies are carried down the river in neutral buoyancy and are in end-to-end contact along the whole length of the pipeline. For a preliminary rough estimate:

(a) Calculate the mean velocity of coal in the pipeline.

(b) Estimate the volumetric flow rate of water needed to transport this coal assuming first laminar flow, then turbulent flow.

(c) Determine whether the flow of water is laminar or turbulent.

(d) Find the theoretical pumping cost per ton to transport the coal assuming a straight run down to Los Angeles in a commercial steel pipe, and 2.5¢/kW hr for electrical energy.

Information for this problem is from *Christian Science Monitor* (June 4, 1982).

2.22. *Plying the Yangtze River.* The Yangtze, now called the Chang-jiang, or the Long River, is China's major transportation artery, allowing goods and passengers to be carried into the heart of the country 2000 km or more from the coast. Innumerable vessels of all types bustle along this waterway, from cruise liners and tankers to barges and junks.

On the lower reaches of the river, between Wuhan and Nanjing, the river divides and reforms again and again. Some of the channels are wide, others are narrow.

The question: When the river splits into two parallel channels, one of length L_1 with flow velocity u_1, the other of length L_2 with flow velocity u_2, which channel would you take if you were fighting your way upstream against the current, and which would you take if you were comfortably cruising downstream?

(a) Develop a general formula to answer this question and then test it by choosing between paths 1 and 2 and between paths 3 and 4.

(b) Now try your formula on paths 5 and 6.

Systems with circulating solids. Knowing how to coax a stream of fine solids to flow by hydraulic action (no screw feed or bucket conveyors) around a loop or back and forth between two process units has led to the development of some of the most important processes in the chemical industry. Fluid catalytic cracking of crude oil is the earliest and probably the most important example of such a process. Finding the solid circulation rate is of prime importance in these operations, and the way this is determined is best illustrated by looking at the air–water analogy.

2.23. Consider the solid circulation loop and a somewhat similar air–water model below. Find the circulation rate of water (kg/s) in the model in which

Air-driven water circulation loop

- $d_t = 2$ cm, $L_{eq,ABC} = 15$ m, $L_{eq,CA} = 1$ m;
- section AC has a vertical rise of 1 m and contains bubbly water, with 1% air, while section ABC is free of air;
- the frictional loss of this 1% air–water mixture is close to that of water alone, with the same critical Reynolds number.

2.24. Consider the circulation of solids between two process vessels and the corresponding air–water model sketched below. Find the circulation rate of water (kg/s) in the pilot plant model in which

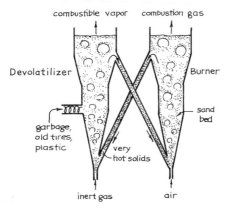

Power-producing municipal incinerator for a city of 100,000 ~ 300,000

Air-driven water circulation system

- for each of the vertical upflow sections AB and CD $d_t = 36$ cm, $L_{eq} = 10$ m, 10% air in bubbly flow;
- for each of the sloping downflow sections BC and DA $d_t = 20$ cm, $L_{eq} = 15$ m, no entrained air;
- for bubbly flow with 10% air assume that the "viscosity" is 1.8 times that of water alone.

Chapter 3   COMPRESSIBLE FLOW OF GASES

When the density change of fluid is small ($\rho_1/\rho_2 < 2$) and the velocity not too high (Mach number, Ma < 0.3), then the mechanical energy balance reduces to the forms developed in Chapter 2. These equations represent the flow of all liquids as well as relatively slow moving gases. This is called *incompressible flow*.

On the other hand, for gases only, when the pressure ratio is large and/or the flow very fast, then kinetic energy and compressibility effects (large density changes) may well become the dominant terms in the mechanical energy balance. In this situation the mechanical energy balance takes on quite different forms from those developed in Chapter 2. This is called *compressible gas flow*, and the equations for this flow are developed in this chapter.

Air flowing through an air conditioning duct can be considered to be incompressible, while flow from a high-pressure tank through a short discharge tube or through a pipeline whose inlet pressure is 50 atm and outlet pressure is 10 atm, should be treated as compressible.

Fig. 3.1

The importance of these two factors, compressibility and velocity, is indicated by the Mach number, defined as

$$
\text{Ma} = \left(\frac{\text{velocity of gas}}{\begin{array}{c}\text{velocity of sound}\\ \text{in the gas}\end{array}} \right)_{\text{at same } T \text{ and } p} = \frac{u}{c} \tag{3.1}
$$

where the speed of sound is given by thermodynamics as

$$
c \underset{\text{gas}}{\overset{\text{ideal}}{=}} \left[\frac{g_c k p}{\rho} \right]^{1/2} = \left[\frac{g_c k R T}{(mw)} \right]^{1/2} \underset{20°C}{\overset{\text{air}}{=}} 343.5 \, \frac{m}{s} \tag{3.2}
$$

Depends on
T, p and ρ

$$k = \frac{C_p}{C_v} \cong 1.67 \quad \text{for monatomic gases}$$
$$\cong 1.40 \quad \text{for diatomic gases}$$
$$\cong 1.32 \quad \text{for triatomic gases}$$

In this chapter we treat in turn

- adiabatic flow in a pipe
- isothermal flow in a pipe
- flow through a nozzle

We then combine the equations for nozzle and pipe to treat the discharge of gases from a tank of reservoir through a pipe, as sketched in Fig. 3.1.

I. ADIABATIC FLOW IN A PIPE WITH FRICTION

Consider a gas flowing through a well-insulated but real pipe, one with frictional resistance. Figure 3.2 shows how conditions change as the gas moves down the pipe.

Suppose p_1 is fixed and p_3 is adjustable. When p_3 is a bit smaller than p_1 flow through the pipe is slow and $p_2 = p_3$. However, as p_3 is lowered more and more the velocity of the gas at the pipe outlet u_2 increases until it

Fig. 3.2. Slow flow in a well-insulated pipe; $p_2 = p_3$.

reaches the speed of sound in that gas. This speed represents the actual mean velocity of motion of the individual molecules of gas.

If p_3 is lowered further the gas leaving the pipe cannot go faster and the velocity remains sonic at p^*, T^*, u^*, and $Ma_2 = 1$; and the flow rate remains unchanged. Thus, the maximum possible flow rate of gas in that pipe has been reached for the particular inlet pressure p_1. So for $p_3 < p^*$ we have what is called *choked flow*, as shown in Fig. 3.3.

To quantitatively characterize this adiabatic flow with friction, use both the mechanical energy and the total energy balances for a differential section of pipe.

$$\underbrace{\frac{g}{g_c} dz}_{\approx 0} + \frac{u\,du}{g_c} + \frac{dp}{\rho} + \underbrace{\Sigma F}_{\frac{2f_F u^2 dL}{g_c d}} + \underbrace{W_s}_{= 0} = 0 \qquad \left[\frac{J}{kg}\right] \qquad (3.3)$$

ρ changes greatly, cannot be assumed to be a constant, and accounts for adiabatic expansion

$$\underbrace{dh}_{C_p dT} + \frac{u\,du}{g_c} + \underbrace{\frac{g}{g_c} dz}_{\substack{\approx 0 \\ = 0 \\ \text{adiabatic}}} = \underbrace{q}_{= 0} - \underbrace{W_s}_{= 0} = 0 \qquad \left[\frac{J}{kg}\right] \qquad (3.4)$$

Fig. 3.3. The fastest possible flow in a well-insulated pipe gives sonic velocity at the exit.

Combining these expressions, writing everything in terms of the Mach number with the help of equations (3.1) and (3.2), assuming a constant value for f_F throughout the length of the pipe, we obtain, after considerable manipulation and integration

$$
\left.
\begin{array}{l}
\dfrac{T_2}{T_1} = \dfrac{Y_1}{Y_2}, \qquad \text{where } Y_i = 1 + \dfrac{k-1}{2}\,\mathrm{Ma}_i^2 \\[3mm]
\dfrac{p_2}{p_1} = \dfrac{\mathrm{Ma}_1}{\mathrm{Ma}_2}\left[\dfrac{Y_1}{Y_2}\right]^{1/2} \\[3mm]
\dfrac{\rho_2}{\rho_1} = \dfrac{\mathrm{Ma}_1}{\mathrm{Ma}_2}\left[\dfrac{Y_2}{Y_1}\right]^{1/2}
\end{array}
\right\} \qquad (3.5)
$$

$$
G = \mathrm{Ma}_1 p_1\left[\dfrac{g_c(mw)k}{RT_1}\right]^{1/2} = \mathrm{Ma}_2 p_2\left[\dfrac{g_c(mw)k}{RT_2}\right]^{1/2} \qquad \left[\dfrac{\mathrm{kg}}{\mathrm{m^2\,s}}\right]
$$

$$
(3.6)
$$

and

$$
\underbrace{\dfrac{k+1}{2}\ln\left[\dfrac{\mathrm{Ma}_2^2 Y_1}{\mathrm{Ma}_1^2 Y_2}\right]}_{\text{KE term}} - \underbrace{\left(\dfrac{1}{\mathrm{Ma}_1^2} - \dfrac{1}{\mathrm{Ma}_2^2}\right)}_{\substack{\text{Compressibility}\\\text{term}}} + \underbrace{k\left(\dfrac{4f_F L}{d}\right)}_{\substack{\text{Pipe}\\\text{resistance}}} = 0 \qquad (3.7)
$$

which relates the flow to the frictional resistance.

The largest possible flow rate in an adiabatic pipe with friction (choked flow) comes with sonic velocity at the exit. Hence, putting $\mathrm{Ma}_2 = 1$ we find

$$
\left.
\begin{array}{l}
\dfrac{T^*}{T_1} = \dfrac{2Y_1}{k+1}\;\underset{k=1.4}{\overset{\text{for }\mathrm{Ma}_1 \approx 0}{=}}\;0.833 \\[3mm]
\dfrac{p^*}{p_1} = \mathrm{Ma}_1\left[\dfrac{2Y_1}{k+1}\right]^{1/2} \\[3mm]
\dfrac{\rho^*}{\rho_1} = \mathrm{Ma}_1\left[\dfrac{k+1}{2Y_1}\right]^{1/2}
\end{array}
\right\} \qquad (3.8)
$$

$$
G^* = \mathrm{Ma}_1 p_1\left[\dfrac{g_c(mw)k}{RT_1}\right]^{1/2} = p^*\left[\dfrac{g_c(mw)k}{RT^*}\right]^{1/2} \qquad \left[\dfrac{\mathrm{kg}}{\mathrm{m^2\,s}}\right] \qquad (3.9)
$$

and

$$\frac{k+1}{2}\ln\left[\frac{2Y_1}{(k+1)\mathrm{Ma}_1^2}\right] - \left(\frac{1}{\mathrm{Ma}^2} - 1\right) + k\left(\frac{4f_\mathrm{F}L}{d}\right) = 0 \quad (3.10)$$

Note that the outlet T, ρ, and p are dependent only on the Mach numbers, and that it is equation (3.7) or (3.10) which relates the Mach numbers with the frictional loss in the length of pipe.

II. ISOTHERMAL FLOW IN A PIPE WITH FRICTION

This represents a long pipeline (see Fig. 3.4). Consider first the flow through a pipe with fixed inlet conditions. As p_3 is lowered the flow rate increases until choking conditions are reached. However, for isothermal flow we will show that the limiting exit Mach number is $1/\sqrt{k}$, instead of 1 which was found for adiabatic flow. So

$$\left.\begin{array}{l} \mathrm{Ma}_{\text{choking, isothermal}} = 1/\sqrt{k} \\[2ex] \mathrm{Ma}_{\text{choking, adiabatic}} = 1 \end{array}\right\} \quad (3.11)$$

Applying the mechanical energy balance to a differential slice of pipe gives

$$\underbrace{\frac{g}{g}dz}_{\cong\,0} + \frac{u\,du}{g_c} + \frac{dp}{\rho} + \Sigma F + \underbrace{W_s}_{=\,0} = 0 \quad \left[\frac{\mathrm{J}}{\mathrm{kg}}\right] \quad (3.12)$$

Ignoring the total energy balance since the gas is isothermal, dividing by u^2,

Fig. 3.4. Slow isothermal flow in a pipe; $p_2 = p_3$.

writing everything in terms of the Mach number, assuming constant f_F, and integrating gives

$$2\ln\frac{Ma_2}{Ma_1} - \frac{1}{k}\left(\frac{1}{Ma_1^2} - \frac{1}{Ma_2^2}\right) + \frac{4f_F L}{d} = 0 \qquad (3.13)$$

$\underbrace{\hphantom{XXXX}}_{\text{KE term}}$ $\underbrace{\hphantom{XXXXX}}_{\substack{\text{Compressibility}\\\text{term}}}$ $\underbrace{\hphantom{XX}}_{\substack{\text{Pipe}\\\text{resistance}}}$

with

$$\left.\begin{aligned} T_2 &= T_1 \\[2mm] \frac{p_2}{p_1} &= \frac{Ma_1}{Ma_2} \\[2mm] \frac{\rho_2}{\rho_1} &= \frac{Ma_1}{Ma_2} \end{aligned}\right\} \qquad (3.14)$$

and

$$G = Ma_1 p_1 \left[\frac{g_c(mw)k}{RT}\right]^{1/2} \qquad \left[\frac{kg}{m^2\,s}\right] \qquad (3.15)$$

For close to isothermal conditions or for high-pressure systems where the gas is nonideal, replace RT by $\bar{Z}R\bar{T}$ in equation (3.15), where $\bar{Z}$ is the mean compressibility factor for the gas in the pipe.

To find the choking conditions (highest possible flow), we go back to the mechanical energy balance of equation (3.12), and on rearranging, write it as:

choking conditions = $-\dfrac{dp}{dL} = \dfrac{2f_F G^2}{g_c \rho d}\left[\dfrac{1}{1-\left(u^2\rho/g_c p\right)}\right] = \dfrac{2f_F G^2}{g_c \rho d}\left[\dfrac{1}{1-kMa^2}\right]$ (3.16)

Now $-(dp/dL)$ can only be positive, never negative; and the highest flow rate results when $-(dp/dL) \to \infty$. Thus, for choked flow, as sketched in

T_1
p_1
ρ_1
$Ma_1 < 1$

$T_{ch} = T_1, p_{ch}, u_{ch}, Ma_{ch} = 1/\sqrt{k}$

surroundings
$p_3 < p_{ch}$

Fig. 3.5. The fastest possible isothermal flow does not give sonic velocity at the exit.

Fig. 3.5, equation (3.16) shows that $Ma_2 = 1/\sqrt{k}$, so

$$\left.\begin{aligned} T_{ch} &= T_1 \\[4pt] p_{ch} &= p_1 Ma_1 \sqrt{k} \\[4pt] \rho_{ch} &= \rho_1 Ma_1 \sqrt{k} \\[4pt] u_{ch} &= \frac{u_1}{Ma_1 \sqrt{k}} \end{aligned}\right\} \qquad (3.17)$$

$$G_{ch} = u_{ch}\rho_{ch} - u_1\rho_1 = G_1 = Ma_1 p_1 \frac{\left(g_c(mw)k\right)^{1/2}}{RT} = p_{ch}\frac{\left(g_c(mw)\right)^{1/2}}{RT}$$

$$(3.18)$$

and

$$\ln\left[\frac{1}{k Ma_1^2}\right] - \left(\frac{1}{k Ma_1^2} - 1\right) + \frac{4 f_F L}{d} = 0 \qquad (3.19)$$

Equation (3.19) relates the frictional resistance to choked flow conditions.

III. WORKING EQUATIONS FOR FLOW IN PIPES

For *adiabatic flow* equation (3.7) in alternative and probably more useful form becomes

$$\frac{k+1}{k}\ln\frac{p_1 T_2}{p_2 T_1} - \frac{k-1}{2k}\left(\frac{p_1^2 T_2^2 - p_2^2 T_1^2}{T_2 - T_1}\right)\left(\frac{1}{p_1^2 T_2} - \frac{1}{p_2^2 T_1}\right) + \frac{4 f_F L}{d} = 0$$

$$(3.20a)$$

with

$$G = \left[\frac{2 g_c(mw)k}{(k-1)R} \cdot \frac{T_2 - T_1}{(T_1/p_1)^2 - (T_2/p_2)^2}\right]^{1/2} \qquad (3.20b)$$

With upstream conditions known and p_2 given, find T_2 from equation (3.20a), then G from equation (3.20b).

Similarly, for *isothermal flow*, equation (3.13) becomes

$$2\ln\frac{p_1}{p_2} - \frac{g_c(mw)}{G^2RT}\left(p_1^2 - p_2^2\right) + \frac{4f_F L}{d} = 0 \qquad (3.21)$$

For situations where nonideal gas behavior and/or somewhat nonisothermal flow occurs we can use the coagulated expression

$$\frac{2}{\alpha}\ln\frac{p_1}{p_2} - \frac{g_c(mw)\left(p_1^2 - p_2^2\right)}{G^2(\overline{Z}R\overline{T})} + \frac{4f_F L}{d} = 0 \qquad (3.22)$$

Deviation from plug flow
$\alpha \cong 1$ for turbulent flow

IV. FLOW THROUGH AN ORIFICE OR NOZZLE

Suppose that gas flows in frictionless adiabatic flow from p_0 to p_1 through a smooth well-rounded orifice as shown in Fig. 3.6. We call this a nozzle. The mechanical and total energy balances then give

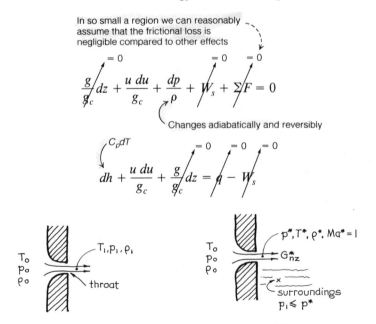

Fig. 3.6. Flow through an orifice or nozzle.   Fig. 3.7. Maximum flow through an orifice.

Accounting for the changing cross-sectional area of the flow channel, and writing everything in terms of the Mach number gives, on integration,

$$
\left.
\begin{aligned}
&\mathrm{Ma}_1^2 = \frac{2}{k-1}(Y_1 - 1) \quad \text{or} \quad Y_1 = 1 + \frac{k-1}{2}\mathrm{Ma}_1^2 \\[2ex]
&\frac{T_1}{T_0} = \frac{1}{Y_1} \\[2ex]
&\frac{p_1}{p_0} = \frac{1}{Y_1^{k/k-1}} \quad \text{or} \quad Y_1 = \left(\frac{p_0}{p_1}\right)^{(k-1)/k} \\[2ex]
&\frac{\rho_1}{\rho_0} = \frac{1}{Y_1^{1/(k-1)}}
\end{aligned}
\right\} \tag{3.23}
$$

$$
G = p_0 \mathrm{Ma}_1 \left[\frac{g_c(mw)k}{RT_0} \frac{1}{Y_1^{(k+1)/(k-1)}} \right]^{1/2} \quad \left[\frac{\mathrm{kg}}{\mathrm{m}^2\,\mathrm{s}} \right] \tag{3.24}
$$

$$
u_1 = \mathrm{Ma}_1 c_1 = \mathrm{Ma}_1 \left[\frac{g_c k R T_0}{(mw)Y_1} \right]^{1/2} \tag{3.25}
$$

If p_1 is lowered there comes a point where the exit gases are at sonic velocity, or $\mathrm{Ma}_1 = 1$. Further lowering of p_1 will not give an increase in flow. For these conditions, sketched in Fig. 3.7,

$K = 1.4$
so for diatomic gases

$$
\left.
\begin{aligned}
&\frac{T^*}{T_0} = \frac{2}{1+k} \xlongequal{k=1.4} 0.833 \\[2ex]
&\frac{p^*}{p_0} = \left(\frac{2}{1+k}\right)^{k/(k-1)} \xlongequal{k=1.4} 0.528 \\[2ex]
&\frac{\rho^*}{\rho_0} = \left(\frac{2}{1+k}\right)^{1/(k-1)} \xlongequal{k=1.4} 0.634
\end{aligned}
\right\} \tag{3.26}
$$

At the throat the maximum flow rate is

$$G_{nz}^* = p_0 \left[\frac{g_c(mw)k}{RT_0} \left(\frac{2}{1+k} \right)^{(k+1)/(k-1)} \right]^{1/2} \underset{k = 1.4}{=\!=\!=} 0.685 \, p_0 \left[\frac{g_c(mw)}{RT_0} \right]^{1/2}$$

$$\left. \underset{20°C,\, 1\,\text{atm}}{\overset{\text{air}}{=\!=\!=}} 239 \frac{\text{kg}}{\text{m}^2\,\text{s}} \right\}$$

$$(3.27)$$

or

$$u^* = C^*\text{Ma}^* = \frac{(2g_c kRT_0)^{1/2}}{(mw)(k+1)} \underset{k = 1.4}{=\!=\!=} 1.080 \left[\frac{g_c RT_0}{(mw)} \right]^{1/2}$$

$$\left. \underset{20°C}{\overset{\text{air}}{=\!=\!=}} 313.6 \frac{\text{m}}{\text{s}} \right\}$$

$$(3.28)$$

A. Comments

1. First check to see whether sonic velocity is reached at the throat, then use the correct equation.

2. By using $p_1/p_0 < 0.5$ we have a simple accurate flow gauge which depends on p_0 and A_{throat} alone; see equation (3.26).

3. For a nonrounded nozzle, say a square orifice, we get all sorts of uncertainties in flow which we don't know how to treat. So we stay away from them. Just round and smooth the opening.

V. PIPE LEADING FROM A STORAGE VESSEL

In a tank–nozzle–pipe system, such as shown in Fig. 3.8, the time needed to go through the nozzle is so short that adiabatic flow is always a

Fig. 3.8. Flow of gas from a storage vessel through a pipe.

good approximation, while flow in the pipe will be somewhere between adiabatic and isothermal—but which? Consider the following:

The pipe flow equations developed in this chapter all assumed one-dimensional (or plug) flow. But this is not strictly correct because a layer of slow-moving fluid always coats the inner wall of the pipe. Consequently, although an adiabatic gas progressively expands and cools as it races down the pipe, it returns to its stagnation temperature whenever it is brought to rest at the pipe wall. Thus, the pipe walls experience the stagnation temperature of the gas despite the fact that the fast-moving gas in the central core may be much cooler.

For this reason, the adiabatic nozzle–adiabatic pipe equations probably better represents the real pipe with constant wall temperature. We consider this situation from now on. In any case, this is not a very serious point because the difference in predictions of the isothermal and adiabatic pipe equations is not significant and can hardly be noted on the performance charts.

Figure 3.9 is a dimensionless plot relating flow rate with overall pressure drop and frictional resistance of the pipe. Note that the longer the pipe the smaller the maximum throughput even though the velocity of gas at the pipe exit can be sonic. Figure 3.10 is a cross-plot of Fig. 3.9, useful for longer pipes.

For long pipes the contribution of the orifice pressure drop becomes negligible and the graphs just represent the equations for pipe flow alone.

REFERENCES

O. Levenspiel, The discharge of gases from a reservoir through a pipe, *AIChE J.* **23**, 402 (1977).

A. H. Shapiro, *The Dynamics and Thermodynamics of Compressible Fluid Flow*, Vol. 1, Chapter 6, Ronald, New York (1953).

A. H. Streeter, *Fluid Mechanics*, Fourth Ed., Chapter 6, McGraw-Hill, New York (1966).

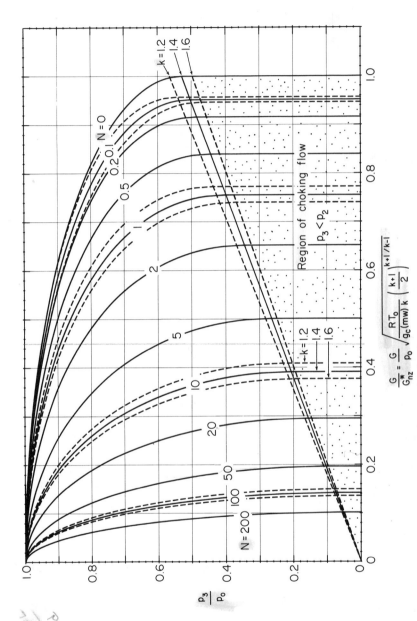

Fig. 3.9. Graph representing the flow of gas through a pipe from a high-pressure storage tank. Simultaneous solution of equations (3.5)–(3.10) with (3.21)–(3.26). [From Levenspiel (1977).]

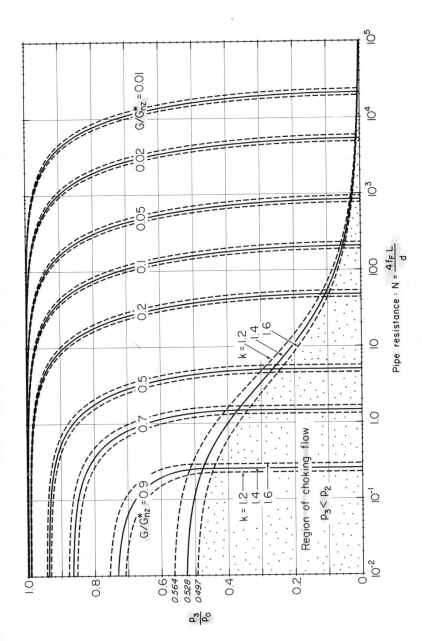

Fig. 3.10. Alternative plot of variables for the tank–nozzle–pipe situation. [From Levenspiel (1977).]

EXAMPLE 3.1. NITROGEN TO AN AMMONIA PLANT

Nitrogen ($k = 1.4$) is to be fed through a 15-mm-i.d. steel pipe 11.5 m long to a synthetic ammonia plant. Calculate the downstream pressure in the line for a flow rate of 1.5 mol/s, an upstream pressure of 600 kPa, and a temperature of 27°C throughout.

Solution

Method A. Let us use equation (3.13). For this, evaluate in turn

$\epsilon = 0.046$ mm (from Table 2.1, pipe roughness)

$$\frac{\epsilon}{d} = \frac{0.046 \text{ mm}}{15 \text{ mm}} = 0.003 \text{ (roughness ratio)}$$

$f_F = 0.00675$ (from Fig. 2.4, assuming that the Reynolds number is high enough; we will check this later.)

$$N = \frac{4f_F L}{d} = \frac{4(0.00675)11.5 \text{ m}}{0.015 \text{ m}} = 20.7$$

$$\dot{m} = \left(1.5 \frac{\text{mol}}{\text{s}}\right)\left(0.028 \frac{\text{kg}}{\text{mol}}\right) = 0.042 \frac{\text{kg}}{\text{s}} \text{ (mass flow rate)}$$

$$G = \frac{\dot{m}}{\text{area}} = \frac{0.042 \text{ kg/s}}{(\pi/4)(0.015)^2 \text{ m}^2} = 237.7 \frac{\text{kg}}{\text{m}^2 \text{ s}} \text{ (mass velocity)}$$

$$\text{Ma}_1 = \frac{G}{p_1}\left[\frac{RT}{g_c(mw)k}\right]^{1/2} = \frac{237.7}{600,000}\left[\frac{8.314(300)}{0.028(1.4)}\right]^{1/2} = 0.1$$

Replacing in equation (3.13) gives

$$-\ln\left(\text{Ma}_1^2/\text{Ma}_2^2\right) - 71.54\left[1 - \left(\text{Ma}_1^2/\text{Ma}_2^2\right)\right] + 20.70 = 0$$

Solve this equation by trial and error.

Guess Ma_1/Ma_2	l.h.s. (left-hand side)
0.5	−31.57
0.4	−37.56
0.8	−4.61
0.87	+3.59
0.84	−0.014 (close enough)
0.841	+0.104

Therefore, from equation (3.14)

$$\frac{p_2}{p_1} = \frac{Ma_1}{Ma_2} = 0.84$$

Thus, the downstream pressure

$$p_2 = 0.84\ p_1 = 0.84(60,000) = 504,000\ Pa$$

Check the Reynolds number to see if it is high enough for f_F to be constant

$$Re = \frac{du\rho}{\mu} = \frac{dG}{\mu} = \frac{(0.015)(237.7)}{1.8 \times 10^{-5}} \cong 200,000$$

This value is just on the borderline, not worth correcting, because it is hard to read the graph any more accurately and also because the data from which this graph is constructed is not that reliable.

Method B. Let us use the recommended working expression for isothermal flow in pipes, or equation (3.21). This can be written as

$$-\ln\left(\frac{p_2^2}{p_1^2}\right) - \frac{g_c(mw)p_1^2}{G^2RT}\left(1 - \frac{p_2^2}{p_1^2}\right) + \frac{4f_F L}{d} = 0$$

and on replacing values

$$-\ln\left(\frac{p_2^2}{p_1^2}\right) - 71.54\left(1 - \frac{p_2^2}{p_1^2}\right) + 20.70 = 0$$

Noting that $p_2/p_1 = Ma_1/Ma_2$, we recognize the above expression as equation (i) of the previous method. The rest follows as with method A.

EXAMPLE 3.2. DESIGN OF A CRITICAL ORIFICE FLOW METER

We want the flow of helium ($k = 1.66$) in a 100-mm-i.d. pipe to be 4 m/s at 105°C and 200 kPa. The storage tank from which we draw helium contains the gas at 1 MPa and -20°C. How do we get the desired flow with a critical orifice?

Solution

The required flow rate is

$$\dot{m} = \left[\frac{\pi}{4}(0.1)^2 4\right]\frac{m^3}{s}\left(\frac{200,000}{101,325}\right)\left(\frac{273}{378}\right)\left(\frac{1\ mol}{0.0224\ m^3}\right)\left(\frac{0.00403\ kg}{mol}\right) = 8 \times 10^{-3}\frac{kg}{s}$$

Since the pressure ratio is well over $2:1$ we have critical flow through the orifice. So equation (3.27) becomes

$$G_{nz}^* = 10^6 \left[\frac{(0.00403)(1.66)}{(8.314)(253)} \left(\frac{2}{2.66} \right)^{2.66/0.66} \right]^{1/2} = 1000 \frac{\text{kg}}{\text{m}^2 \text{ s}}$$

To find the diameter of hole needed note that

$$\dot{m} = AG = (\pi/4) d^2 G$$

or

$$d = \left(\frac{4}{\pi} \frac{\dot{m}}{G} \right)^{1/2} = \left(\frac{4}{\pi} \frac{8 \times 10^{-3}}{1000} \right)^{1/2} = \boxed{3.19 \text{ mm}}$$

Comment. Since we end up with a small-diameter orifice followed by a large-diameter pipe it is safe to ignore the resistance of the pipe.

EXAMPLE 3.3. USE OF DESIGN CHARTS FOR FLOW OF GASES

Air at 1 MPa and 20°C in a large tank discharges to the atmosphere through 12.5 m of 15-mm-i.d. drawn tubing. What is the pressure half-way down the tubing?

Solution

Since this problem involves a tank–nozzle–pipe we can solve it with the design charts of this chapter. So referring to the drawing below let us evaluate the terms needed to use the charts.

$$\epsilon = 0.0015 \text{ mm (from Table 2.1)}$$

$$\frac{\epsilon}{d} = \frac{0.0015 \text{ mm}}{15 \text{ mm}} = 0.0001$$

$$f_F = 0.0003 \text{ (from Fig. 2.4)} \qquad ,003$$

$$N = \frac{4f_F L}{d} = \frac{4(0.003)12.5 \text{ m}}{0.015 \text{ m}} \cong 10$$

$$\frac{p_3}{p_0} = \frac{101,325}{10^6} \cong 0.1$$

With this information, locate point A in Fig. 3.9, as shown in the sketch below.

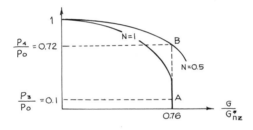

This gives $G/G_{nz}^* = 0.76.$ $.39$

Half-way down the pipe G/G_{nz}^* remains unchanged at 0.76, however $N = 0.5$. This locates point B which in turn gives

$$\frac{p_4}{p_0} = 0.72 \qquad N = 5$$

Thus, the pressure half-way down the pipe is

$$p_4 = 720 \text{ kPa}$$

PROBLEMS ON THE FLOW OF GASES

3.1. It bugs me that they are so tightlipped about the production rate of their new coal gasification plant. But, perhaps we could work it out for ourselves. I noted that the gas produced ($mw = 0.013$, $\mu = 10^{-5}$ kg/m s, $k = 1.36$) is sent to neighboring industrial users through a bare 15-cm-i.d. pipe 100 m long. The pressure gauge at one end of the pipe read 1 MPa absolute. At the other end it read 500 kPa. I felt pain when I touched the pipe, but when I spat on it, it didn't sizzle so I guess that the temperature is 87°C. Would you estimate for me the flow rate of coal gas through the pipe, both in tons/day and in m^3/s measured at 1 atm and 0°C?

3.2. For our project on the biochemical breakdown of grass straw we need to oxygenate the deep fermentation vat by introducing 5 lit/s of air at 2 atm. We obtain this air from a compressor located 100 m away through a steel pipe 0.1-m-i.d. What should be the pressure at the pipe inlet to guarantee this flow rate?

3.3. Repeat Example 3.1 with one change; the flow rate is to be doubled to 3 mol/s.

3.4. *Hydrogen pipelines.* If the United States converts to a hydrogen economy the electricity produced by power plants and dams will be used directly to decompose water into its elements. Then, instead of pumping electricity everywhere on unsightly power lines, hydrogen will be pumped in a network of unseen underground pipelines. This hydrogen will be used to fuel autos, to heat homes, and to produce small amounts of needed electricity. If this day comes, one trunk pipeline (0.5-m-i.d.) will come direct to Corvallis from Bonneville Dam 300 km away. The pressure of H_2 entering the pipeline will be 2 MPa; at Corvallis, it will be 1 MPa. The temperature throughout the line can be estimated to be 20°C. Find the flow rate of hydrogen under these conditions in kg/s and in std. m^3/s (at 1 atm, 0°C).

Consider various aspects of the proposed hydrogen pipeline of the above problem.

3.5. If gas consumption at Corvallis rises high enough so that the pipeline pressure at Corvallis drops from 1 to 0.5 MPa, what would be the gas flow rate? The pressure at Bonneville would remain unchanged.

3.6. If a 1-m pipeline were laid instead of the 0.5-m line, how much gas in kg/s and in std. m^3/s could be transported?

3.7. If the hydrogen pressure at Bonneville were doubled to 4 MPa, while the pressure at Corvallis stays at 1 MPa, what flow rate can be obtained?

3.8. Nitrogen ($k = 1.39$) at 200 kPa and 300 K flows from a large tank through a nozzle (throat diameter = 0.05 m) into surroundings at 140 kPa. Find the mass flow rate of nitrogen and compare it with choked flow. [This problem was taken from F. A. Holland, *Fluid Flow for Chemical Engineers*, p. 125, Arnold, London (1973).]

3.9. A safety vent from a 3.5-atm reactor to the atmosphere consists of 5 m of 15-mm commercial steel pipe leading off from the reactor. Is flow choked or not?

3.10. Methane ($k = 1.2$) discharges from one tank (112 kPa) to another (101 kPa) through 2.4 m of 7.66-mm-i.d. pipe. I am afraid that the connecting pipe will snap off at the upstream end since it is just held in place there with chewing gum. If it does, what will happen to the discharge rate of methane from the upstream tank?

3.11. 25 mol/s of ethylene ($k = 1.2$, $\mu = 2 \times 10^{-5}$ kg/m s) are to be fed to a reactor operating at 250 kPa from a storage tank at 60°C and 750 kPa. This flow is to be controlled by a discharge control tube 24-mm-i.d. made of commercial iron pipe as shown below. What length of control tube is needed?

3.12. Hydrogen ($k = 1.4$) flows from a tank at 1 MPa to a second tank at 400 kPa through 36 m of 115-mm commercial steel pipe. What is the pressure p_1 in this connecting pipe just outside the high-pressure tank?

3.13. Here's my flow meter—2 m of 3-cm pipe sticking out from the gas storage tank (-20°C), ending with a cap having a 0.5-mm hole drilled in it so as to make a critical orifice. However, the size of hole is not right, because the flow is exactly 8% too high. Even if I drill another hole, I'll most likely still be off, so no more holes for me. What should I do? How to get the right flow rate with the present orifice? Make your calculations and present your answer with a sketch.

3.14. Drat! The hole drilled into the wall of the tank (5 atm inside, 1 atm outside) is too big because the flow is 5 times that desired. With a 200 mm long pipe fitted to the hole things are better since the flow is only 2.5 times that desired. What size of pipe should we fit to the hole to get just the right flow rate?

3.15. With a hole at the wall of the storage tank (200 kPa inside) the discharge to the surroundings (144 kPa) is 2.5 times that desired. But when we attach a tube 1 m long to the orifice (same diameters) the flow rate is three-fourths that desired. What length of tube should we use to get the right flow rate?

3.16. When the pressure in the tank rises to 180 kPa the relief valve to the atmosphere (100 kPa) opens. When the pressure drops to 110 kPa the valve closes and the pressure rises again. Estimate the ratio of flow rates of air at the end and beginning of this operation if the temperature remains at 300 K throughout the cycle.

3.17. Air vents from a reactor (0°C, 982 kPa) through a smooth pipe (d = 3.75 mm, L = 10.4 m) into the atmosphere. Find the flow rate of air in mol/s.

3.18. Repeat the previous problem for a pipe length of 2.08 m instead of 10.4 m.

3.19. A 23-mm steel safety vent pipe 19.6 m long leads from the air storage tank (167 kPa) to the atmosphere. But flow is too low in the vent pipe. What should be the length of this vent pipe if the air flow is to be 80% higher than in the present pipe?

Oxygen is fed to a reactor operating at 240 kPa from a storage tank at 300 kPa through 1187.5 mm of 7.6-mm-i.d. steel pipe. The flow rate is not high enough. By what percentage or fraction does the flow rate change

3.20. if the pressure of oxygen in the storage tank is raised to 800 kPa?

3.21. if the original connecting pipe is replaced by one which is twice the diameter?

Gas flows from tank A through a pipe to tank B, and the pressures in the tanks are such that choked flow exists in the pipe. What happens to the flow rate

3.22. if the pressure in tank A is doubled and the pressure in tank B is halved?

3.23. if the pressure in tanks A and B are both doubled?

3.24. Let us see whether the adiabatic assumption or the isothermal assumption for a pipe gives the more conservative estimate of flow rate. Compare the adiabatic and isothermal flow rates of air in pipes which are 10-mm-i.d. and 1 m long, with identical upstream conditions of 1 MPa and 300 K, and with exit pressure low enough so that choked or critical flow exists in these pipes.

Fig. 3.11. The heat pipe with an adiabatic section.

Information on Heat Pipes, Useful for Problems 3.25 to 3.27

The heat pipe is a crafty and efficient way of transferring heat from a hot place to a cold place even with a very small temperature difference. It consists of a sealed pipe with a wick going from end to end and containing just the right fluid. At the hot end (the evaporator) the vapor pressure is high, so liquid boils. At the cold end the vapor pressure is low, so vapor flows to that end and condenses. Then, by capillarity, the condensed liquid is drawn along the wick from the cold end back to the hot end. As an illustration consider the heat pipe containing water and steam shown in Fig. 3.11.

The capacity of a heat pipe is enormous and is limited by one of the following factors:

- Heat transfer rates into and out of the pipe at the hot and cold ends.
- The capillary stream may break, meaning that the hot end uses up liquid faster than the wick can suck it from the cold end.
- The vapor flow from hot to cold end under the prevailing pressure difference may be the slow step.

Here we only consider the last factor. To estimate this limiting flow note that vapor speeds up in the evaporator section, moves fastest in the adiabatic section, then slows down in the condenser. As an approximation we can look at this as an orifice followed by a pipe (the adiabatic section).

Sometimes there is no adiabatic section. In this situation vapor speeds up to a maximum at the boundary between evaporator and condenser, then slows down. This can be looked upon as a nozzle, or an orifice, as shown in Fig. 3.12.

Fig. 3.12. The heat pipe with no adiabatic section.

3.25. A *water heat pipe* 1.5 m long is to be used to help equalize the temperature in two side-by-side regions. The hot portion of the pipe is 0.6 m long and contains boiling water at 124°C (vapor pressure = 225 kPa); the cool portion is 0.9 m long and contains condensing steam at 90°C (vapor pressure = 70 kPa). Find the heat transfer rate if the diameter of the vapor transport section is 1 cm, and if vapor transport is the limiting process. Each kilogram of water going from hot to cold end transfers 2335 kJ of heat.

3.26. *Cryogenic medical probe.* The compact hand-held cryogenic probe sketched below is designed to freeze tumors and tissues. It consists of a small high-pressure nitrogen heat pipe having an evaporator section (fluid at 100 K and 799 kPa) to contact the tumor, a 0.3 m adiabatic section, and a condenser (fluid at 85 K and 229 kPa) bathed in liquid nitrogen and open to the atmosphere. Estimate the heat removal rate of the tip if the inside diameter of the heat pipe is 2 mm, its roughness is 0.02 mm, and if vapor transport is limiting. Each kg of boiling and condensing nitrogen transports 193 kJ of heat.

3.27. *Heat pipes for a solar-heated home.* In one design a set of ammonia heat pipes is used to transport heat from the solar-heated reservoir located in the basement of the home to its living quarters. One such pipe consists of an evaporator section immersed in the heat reservoir (fluid at 31°C, 1200 kPa), a long adiabatic section (10 m), and a condenser (fluid at 28°C, 1100 kPa). Find the heat transfer rate of this heat pipe if the vapor transport section has a diameter of 16 mm, a surface roughness $\epsilon = 0.096$ mm, and if vapor transport is limiting. Each kilogram of ammonia going from the hot end to the cold end transports 1155 kJ of heat.

3.28. *Space satellite heat pipe.* An ammonia heat pipe is to transfer heat from the hot side to the cold side of a small space satellite. The large evaporator and condenser sections of this unit are connected by an adiabatic vapor transport section 5-mm-i.d., surface roughness of 0.075 mm, and 0.84 m long. At the hot end of the heat pipe ammonia boils at 25°C ($\pi = 1$ MPa, $H_{vapor} = 1465$ kJ/kg), at the cold end ammonia condenses at -2°C ($\pi = 400$ kPa, $H_{liquid} = 172$ kJ/kg). Find the heat transfer rate of this unit if vapor transport is limiting. Consider the average temperature of the adiabatic section to be 12°C. [Problem modified from P. D. Dunn and D. A. Reay, *Heat Pipes*, second ed., p. 133, Pergamon Press, New York (1978).]

3.29. *Accidents at nuclear power stations.* The reactor core of a nuclear power station is immersed in a pool of water and it has all sorts of safety devices to guard against the following two types of accidents:

- Inadequate heat removal. This would result in a rise in temperature and pressure, with possible rupture of the container.
- Loss of coolant. This would expose the reactor core, cause a meltdown, and destroy the unit.

The last-ditch safety device, which should never ever have to be used, is a water standpipe 10 m high, 1-m-i.d. having a safety vent 0.35 m in diameter on top, and emergency water input from below, as shown below.

The inconceivable has happened. Elsewhere in the plant something went wrong, safety devices didn't work, the net result being that the heat exchanger fluid stopped removing heat from the reactor. Within 8 s the moderator rods fell into place, shutting down the reactor; however, the residual heat release caused the temperature and pressure to build up and thereby blow open the safety valve on the standpipe. Water boils madly and steam screams out the top of the standpipe, as shown here.

You are standing by your sightglass, hand close to the emergency water button, but so far all is fine, the water level in the sightglass shows 7 m. However, you wonder whether the high steam velocity in any way can falsify your reading and whether the water level in the standpipe really is at 7 m. If not, what is the water level? With a pencil and calculator handy, with the following properties of water known and with one eye on the sightglass, please make the necessary calculations.

Data: For water at 15 MPa and 342°C

$$\rho_{\text{liquid H}_2\text{O}} = 603 \text{ kg/m}^3$$

$$\rho_{\text{steam}} = 96.7 \text{ kg/m}^3$$

$$k = C_p/C_v = 1.26$$

[Problem adapted from Darrel Teegarden. Also see J. G. Collier and L. M. Davies, "The Accident at Three Mile Island," *Heat Transfer Eng.* **1**, 56 (1980).]

NOTE: If you don't have your answer within 82 seconds—run for it!

Chapter 4   MOLECULAR FLOW

The mean free path (mfp) of molecules increases when the gas pressure is reduced, and at low enough pressure the (mfp) is so large that the molecules begin to bounce from wall to wall of the flow channel rather than collide with each other. When this happens the character of the flow changes. Thus, different flow regimes are encountered depending on the value of the ratio

$$Kn = \frac{(mfp)}{d} = \frac{\text{mean free path of molecules}}{\text{diameter of flow channel}}$$

where Kn is called the Knudsen number. These flow regimes are as follows:

- *Ordinary laminar flow* (Kn $\ll$ 1): Here Poiseuille's law applies. Flow in this regime is based on the following two assumptions
  (a) $\tau = (\mu/g_c)(du/dy)$ with μ = const
  (b) velocity at the wall is zero.
- *Intermediate or slip flow regime* (Kn $\cong$ 1): Here assumption (b) begins to break down.

Fig. 4.1. Velocity profiles in various flow regimes.

- *Molecular flow regime* (Kn ≪ 1): Here there are very few collisions between molecules. Most collisions are with the wall. So the concept of viscosity has no meaning and assumption (a) also breaks down.

The velocity profiles in these three regimes are shown in Fig. 4.1.

Now the mean free path of gas molecules varies with pressure, and from kinetic theory of gases is found to be about as follows:

$$\text{at 1 atm:}\quad (mfp) = 6.8 \times 10^{-8} m$$

$$\text{at 1 Pa:}\quad (mfp) = 6.8 \times 10^{-3} m$$

From this we have

$$\left.\begin{array}{ll}
\text{Laminar flow:} & \text{when } pd > 0.8 \text{ Pa m} \\[2mm]
\text{Molecular flow:} & \text{when } pd < 0.01 \text{ Pa m}
\end{array}\right\} \tag{4.1}$$

In finding how fluids flow in high-vacuum systems we may have to consider all three flow regimes. In just about all cases our concern reduces to handling the tank–line–pump problem, such as shown in Fig. 4.2. At one end of the system flow may be in one regime, at the other end in another. We look at such problems. Also, in evacuating a system the pressure decreases with time, consequently the mass velocity decreases, and the

Fig. 4.2. Typical problem for vacuum systems.

Reynolds number becomes very small. Thus, the transition normally is from laminar to molecular flow, rarely from turbulent to molecular flow.

I. EQUATIONS FOR FLOW, CONDUCTANCE, AND PUMPING SPEED

A. Notation

Molecular flow has its own particular and convenient notation. Let us introduce three essential terms.

1. Flow rate. This is measured by

$$Q = \left(\dfrac{\text{m}^3 \text{ of gas flowing if the pressure is corrected to unit pressure, or 1 Pa}}{\text{time}} \right)$$

$$= p\dot{v} = \dot{n}RT = \frac{\dot{m}RT}{(mw)} = p\frac{\pi d^2}{4}u = \frac{\pi d^2}{4}\frac{GRT}{(mw)}$$

$$= \frac{\pi d}{4}\frac{RT\mu}{(mw)}(\text{Re}) \quad \left[\frac{\text{Pa m}^3}{\text{s}} = \frac{\text{N m}}{\text{s}} = \text{W} \right] \tag{4.2}$$

2. Conductance. In a flow channel such as sketched in Fig. 4.3, the flow rate is proportional to the driving force, Δp. Thus

$$Q = - C_{12}\Delta p = C_{12}(p_1 - p_2) \tag{4.3}$$

$$\text{m}^3/\text{s}$$

where C_{12} is called the conductance between points 1 and 2, and is inversely

Fig. 4.3   Fig. 4.4   Fig. 4.5

proportional to the resistance to flow in that section of flow channel, or

$$C_{12} \propto \frac{1}{\text{resistance}}$$

3. Pumping speed. The volumetric flow rate of material across a plane normal to flow is called the pumping speed S. Thus, at planes A and B of Fig. 4.4, we have

$$Q = S_A p_A = \underset{\underset{m^3/s}{\big\uparrow}}{S_B p_B} \tag{4.4}$$

Note the distinction between pumping speed and conductance. Although they have the same dimensions (m^3/s) they are different measures and should not be confused. C refers to a section of flow system, while S measures what passes across a plane normal to flow. Thus, in Fig. 4.5, C_{12} refers to section 1–2, S_A refers to plane A. The following sections will present equations for conductances, pumping speed, and flow rates for various sorts of equipment: pipes, orifices, pumps, and fittings.

B. Laminar Flow in Pipes

In any differential section of pipe in which there is isothermal laminar flow, the mechanical energy balance of Chapter 1 becomes

$$\cancel{\frac{g}{g_c} dz} + \cancel{\frac{u\,du}{g_c}} + \frac{dp}{\rho} + \cancel{W_s} + \Sigma F = 0$$

Ignore   Ignore   = 0, $\dfrac{32u\mu}{g_c d^2 \rho}\,dL$ [from equation (2.5)]
                          no pump

Integrating and combining with equation (4.3) gives

$$Q_{\text{lam}} = C_{\text{lam}}(p_1 - p_2) \qquad \left[\frac{\text{Pa m}^3}{\text{s}}\right]$$

where

$$\bar{p} = \frac{p_1 + p_2}{2}$$

$$C_{\text{lam}} = \frac{\pi g_c d^4 \bar{p}}{128\,\mu L} \underset{20°C}{\overset{\text{air}}{=\!=\!=}} 1364 \frac{d^4 \bar{p}}{L}$$

$$\underset{20°C}{\overset{\text{H}_2\text{O vapor}}{=\!=\!=}} 2584 \frac{d^4 \bar{p}}{L} \qquad \left[\frac{\text{m}^3}{\text{s}}\right]$$

$$\tag{4.5}$$

Fig. 4.6. Two types of collisions of molecules with the pipe wall: (a) elastic collisions; (b) diffuse reflections.

C. Molecular Flow in Pipes

In this regime we assume no collision between molecules, they simply float from wall to wall of the pipe. But how do molecules leave the wall? Do they bounce off the wall (elastic collision) as shown in Fig. 4.6a, or do they hesitate for a long enough time on the surface to forget the direction they originally came from (diffuse reflection) as shown in Fig. 4.6b?

Let f = fraction of molecules diffusely reflected. For these Knudsen showed that the number leaving at any particular angle is given by

$$n = k \cos \theta$$

Then $1 - f$ = fraction reflected or bouncing off the wall.

Very little information is available on the value of f, but roughly

$$f \cong 0.77 \text{ for copper and glass tubing}$$

$$f \cong 0.90 \text{ for iron pipe}$$

Also, f values are suspected to vary with flow regime; for example, see Fig. 4.7.

Because of the uncertainty in f value and because it is close to unity we will assume throughout that $f = 1$. Then, on applying the kinetic theory of gases with this assumption, it can be shown that

$$Q_{\text{mol}} = C_{\text{mol}}(p_1 - p_2) \qquad \left[\frac{\text{Pa m}^3}{\text{s}} \right]$$

Fig. 4.7

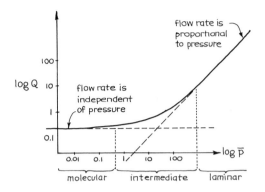

Fig. 4.8. Flow rate of a gas in a pipe for a fixed Δp between the two ends.

where

$$C_{mol} = \frac{d^3}{L}\left[\frac{\pi g_c RT}{18(mw)}\right]^{1/2} \quad \frac{air}{20°C} \quad 121.3\frac{d^3}{L}$$

$$\left.\frac{H_2O \text{ vapor}}{20°C}153.7\frac{d^3}{L}\quad\left[\frac{m^3}{s}\right]\right\} \quad (4.6)$$

D. Intermediate or Slip Flow Regime

If we simply add the laminar and molecular contributions to the total flow as the pressure shifts from one flow regime to the other we find the behavior shown in Figs. 4.8 and 4.9. Actually, the observed flow in the slip flow regime is somewhat lower (at most 20%) than the sum of the individual

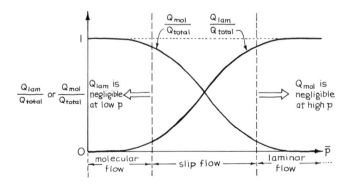

Fig. 4.9. Relative contribution of laminar and molecular mechanisms to the flow of gases in pipes.

contributions. Since the more exact treatment of this situation would lead to complications we will assume simply that

$$Q_{\substack{\text{total in} \\ \text{slip flow}}} = Q_{\text{mol}} + Q_{\text{lam}} \qquad \left[\frac{\text{Pa m}^3}{\text{s}} \right] \qquad (4.7)$$

More precise equations for this flow regime are found in Dushman (1949).

E. Orifice, Contraction, or Entrance Effect in the Molecular Flow Regime

As shown in Fig. 4.10 we have two situations here: an orifice or obstruction in a length of pipe (case A), and a smaller pipe leading from a larger pipe or a tank (case B). For both cases the kinetic theory of gases gives

$$Q_{\text{or,mol}} = C_{\text{or,mol}}(p_1 - p_2) \qquad \left[\frac{\text{Pa m}^3}{\text{s}} \right]$$

where

$$C_{\text{or,mol}} = d^2 \left(\frac{D^2}{D^2 - d^2} \right) \left[\frac{\pi g_c RT}{32(mw)} \right]^{1/2}$$

$$\underset{20°C}{\overset{\text{air}}{=\!=\!=}} 91 d^2 \left(\frac{D^2}{D^2 - d^2} \right)$$

$$\underset{20°C}{\overset{\substack{\text{water} \\ \text{vapor}}}{=\!=\!=}} 115 d^2 \left(\frac{D^2}{D^2 - d^2} \right) \qquad \left[\frac{\text{m}^3}{\text{s}} \right]$$

$$(4.8)$$

1. Equivalent length of pipe. Comparing conductances with flow in a tube, or equation (4.8) vs. equation (4.6), shows that the length of pipe

Fig. 4.10

which has the same resistance as the orifice or contraction is:

- For case A, in terms of a pipe of diameter D,

$$\frac{L_{eq}}{d} = \frac{4}{3}\left(\frac{D^2}{d^2} - 1\right) \tag{4.9}$$

This value of L_{eq} can be quite large. Thus, if $d = 0.1D$, then the resistance of the orifice is equivalent to the resistance of a pipe which is 132 pipe diameters long.

- For case B, in terms of the smaller following pipe of diameter d,

$$\frac{L_{eq}}{d} = \frac{4}{3}\left(1 - \frac{d^2}{D^2}\right) \tag{4.10}$$

This expression shows that the resistance contributed by the contraction is equivalent to a length of about one diameter of small pipe. This is often negligible compared to the other resistances in the vacuum system.

F. Contraction in the Laminar Flow Regime

Consider laminar flow of gases, not molecular flow, at not too high velocities (not near critical flow) through a contraction going from D to d. From the values of Table 2.2 we can show that

$$\left. \begin{array}{c} Q_{or,lam} = C_{or,lam}(p_1 - p_2) \\[2ex] C_{or,lam} = \dfrac{\pi g_c \bar{p}}{\rho u}\left(\dfrac{d^2 D^2}{D^2 - 0.8 d^2}\right) \end{array} \right\} \tag{4.11}$$

where

The equivalent length of this contraction, in terms of the leaving pipe d, is then found to be

$$\frac{L_{eq}}{d} = \frac{Re}{160}\left(1.25 - \frac{d^2}{D^2}\right) \tag{4.12}$$

Here the equivalent length can be as much as 18 diameters of small pipe.

G. Critical Flow Through a Contraction

When the pressure ratio across a contraction is ≥ 2 the contraction behaves as a critical flow orifice. For this situation equation (3.27) can be

written as

$$Q = \frac{\pi}{4} d^2 P_{\text{upstream}} \left[\frac{kRT}{(mw)} \left(\frac{2}{1+k} \right)^{(k+1)/(k-1)} \right]^{1/2} \qquad (4.13)$$

H. Small Leak in a Vacuum System

Suppose we have a tiny leak in a vacuum system. One may look at this in one of a number of ways; for example, as a narrow channel or as a pinch point. These two extremes are shown in Fig. 4.11. Let us estimate the leak rate for these two extremes, remembering that $P_{\text{system}} \ll P_{\text{surroundings}}$.

1. Assume a capillary. If the diameter of the capillary is small compared to the mean free path of molecules at 1 atm, then we have molecular flow of the leaking gas, and from equation (4.6),

$$Q_{\text{leak}} = Q_{\text{mol}} = \frac{d^3 P_{\text{upstream}}}{L} \left[\frac{\pi g_c RT}{18(mw)} \right]^{1/2} \qquad (4.14)$$

However, if the diameter of the capillary is large compared to the mean free path of the molecules at 1 atm, then we have laminar flow of the leaking gas most of the way through the capillary (see problem 4.3), in which case equation (4.5) applies. This gives

$$Q_{\text{leak}} = Q_{\text{lam}} = \frac{\pi g_c d^4 \left(P_{\text{upstream}}^2 - P_{\text{downstream}}^2 \right)}{256 \, \mu L} \qquad (4.15)$$

2. Assume an orifice. Since the pressure ratio across the orifice is many times greater than 2 we should use the critical orifice expressions for compressible flow. Thus, from equation (3.27) we have

$$Q_{\text{leak}} = Q_{\text{crit}} = \frac{\dot{m}RT}{(mw)} = \frac{G^* ART}{(mw)}$$

$$= \frac{\pi}{4} d^2 P_{\text{upstream}} \left[\frac{kRT}{(mw)} \left(\frac{2}{1+k} \right)^{(k+1)/(k-1)} \right]^{1/2} \qquad (4.16)$$

here we have a narrow channel, assumed to be a thin capillary

here we have a pinch point, assumed to be an orifice

Fig. 4.11. Two ways of looking at a leak in a vacuum system.

Which equation you use, (4.14), (4.15), or (4.16), depends on what you know of the leak and how you view it. However, if you know nothing of the nature of the leak, assume the critical orifice. Chances are that this extreme more closely represents the leak.

I. Elbows and Valves

In the molecular flow regime and Re < 100 the resistance of elbows and valves which have no flow restrictions is negligible. So just take the mean flow length of the fitting, bend, open valve, etc. However, if the pipe fitting or valve has a restriction, find the smallest cross section and apply equation (4.9).

J. Pumps

We define pumping speed S_p as follows

$$S_{p,\,\text{at } p_1} = \left(\begin{array}{c} \text{volume of gas removed,} \\ \text{measured at } p_1 \end{array} \middle/ \text{time} \right)$$

$$= \left(\begin{array}{c} \text{volume of gas entering the throat} \\ \text{of the pump, measured at the} \\ \text{entrance of the pump} \end{array} \middle/ \text{time} \right)$$

$$= \frac{Q}{p_1} \qquad \left[\frac{\text{m}^3}{\text{s}} \right] \tag{4.17}$$

The maximum theoretical pumping speed can be looked upon as the flow rate into an orifice which has no back pressure or, with equation (4.16)

$$S_{p,\,\text{max}} = \frac{Q_{\text{or}}}{p_1} \frac{\text{air}}{20°\text{C}} 91d^2 \tag{4.18}$$

The speed factor of a pump is defined as follows:

$$\text{Speed factor} = \left(\frac{\text{speed of actual pump}}{\text{speed of a perfect vacuum pump}} \right) \quad (4.19)$$

At pressures between 10^{-4} and 1 Pa the speed factor is equal to 0.4 ~ 0.6 for an oil diffusion pump; and is equal to 0.1 ~ 0.2 for a mercury vapor pump. The maximum practical speed factor of vacuum pumps $\cong 0.4$.

II. CALCULATION METHOD FOR PIPING SYSTEMS

Suppose we have the piping system shown in Fig. 4.12. Its flow resistances, included in this figure, consists of a rather complex series–parallel combination. Let us see how to evaluate the overall resistance to flow for systems of this kind.

1. For *resistances in parallel*, an example being molecular and laminar flow, we write

$$Q_{\text{mol}} = C_{\text{mol}}(p_1 - p_2)$$

$$Q_{\text{lam}} = C_{\text{lam}}(p_1 - p_2)$$

Adding these flow contributions gives

$$(Q_{\text{mol}} + Q_{\text{lam}}) = (C_{\text{mol}} + C_{\text{lam}})(p_1 - p_2)$$

or

$$Q_{\text{tot}} = C_{\text{tot}}(p_1 - p_2) \quad (4.20)$$

Fig. 4.12. A piping system and its corresponding series–parallel resistances.

2. For *resistances in series*, noting that the flow rate is the same for each section, we can write

$$-Q = C_{12}\,\Delta p_{12} = C_{23}\,\Delta p_{23}$$

or

$$Q = C_{12}(p_1 - p_2) = C_{23}(p_2 - p_3)$$

Eliminating the intermediate pressure p_2 we find

$$Q = \frac{1}{(1/C_{12}) + (1/C_{23})}(p_1 - p_3) = C_{\text{tot}}(p_1 - p_3) \qquad (4.21)$$

Extending this procedure to any number of regions in series is direct, and generalization to any arrangement of series and parallel resistance is not too difficult. For example, for the tank–line–pump system of Fig. 4.13, a frequently met situation, we have

$$Q = C_{01}(p_0 - p_1) = C_{12}(p_1 - p_2) = S_2\,p_2$$

Combining these expressions and eliminating intermediate partial pressures p_1 and p_2 gives

Fictitious total conductance between 0 and 3 where $p_3 = 0$.
It is in fact the pumping speed at the tank, or S_0.

$$Q = C_{\text{tot}}(p_0 - p_3) = S_0\,p_0$$

$$= \frac{1}{\dfrac{1}{C_{\text{or}}} + \dfrac{1}{C_{\text{line}}} + \dfrac{1}{S_p}}\,p_0$$

$$= \frac{1}{\dfrac{1}{C_{01,\,\text{lam}} + C_{01,\,\text{mol}}} + \dfrac{1}{C_{12,\,\text{lam}} + C_{12,\,\text{mol}}} + \dfrac{1}{S_2}}\,p_0 \qquad (4.22)$$

$$\begin{array}{c}\text{air}\\ \overline{\overline{20°C}}\end{array}\quad \frac{1}{\dfrac{1}{\dfrac{\pi g_c p d^2}{\rho u} + 91d^2} + \dfrac{1}{1364d^4\bar{p} + 121.3d^3} + \dfrac{1}{S_p}}\,p_0$$

Eq. (4.11)   Eq. (4.8)   Eq. (4.5)   Eq. (4.6)

These two terms can usually be ignored

$$= \frac{p_1 + p_2}{2},$$ and for anything but the shortest pipes take $p_1 \cong p_0$

Fig. 4.13

look at the pump as a
sink with pressure p = 0

III. PUMPING DOWN A VACUUM SYSTEM

Consider the changing conditions where gas is being pumped from the system, gas leaks into the system, and the pressure within the system

$P_0 \to P_{0,\infty}$ as $t \to \infty$

Q leak

decreases with time. At any instant we can express the net flow rate of gas out of the system as

$$Q = -\frac{d}{dt}(p_0 V) = -V\frac{dp_0}{dt} = \left(\begin{array}{c}\text{removal rate}\\\text{from tank}\end{array}\right) - \left(\begin{array}{c}\text{leak rate}\\\text{into tank}\end{array}\right)$$

$$= S_2 p_2 - Q_{\text{leak}} \qquad\qquad\qquad (4.23)$$

$$= C_{\text{tot}} p_0 - Q_{\text{leak}}$$

At $t = \infty$ the system reaches steady state, $Q = 0$, so the above expression reduces to

$$Q_{\text{leak}} = C_{\text{tot}} P_{0,\infty} \qquad\qquad (4.24)$$

Combining equations (4.23) and (4.24) at any time gives

$$-\frac{dp_0}{dt} = \frac{C_{\text{tot}}}{V}(p_0 - P_{0,\infty}) \qquad Q_{\text{leak}}/C_0$$

where $\qquad\qquad\qquad\qquad\qquad\qquad\qquad\qquad (4.25)$

$$\frac{1}{C_{\text{tot}}} = \frac{1}{C_{\text{or}}} + \frac{1}{C_{\text{line}}} + \frac{1}{S_p}$$

In general C_{tot} varies with pressure.

Now if C_{tot} is constant and independent of p (this means that C_{line} and S_p do not change as the pressure is lowered) integration gives

$$\ln\left(\frac{p_{0,start} - p_{0,\infty}}{p_{0,finish} - p_{0,\infty}}\right) = \frac{C_{tot}}{V} t \qquad (4.26)$$

If unknown, start by ignoring this

If C_{tot} is not constant but varies with p (usually S_p varies drastically with pressure) integration gives

$$-\int_{p_{0,start}}^{p_{0,finish}} \frac{dp}{C_{tot}(p_0 - p_{0,\infty})} = \frac{1}{V}\int_0^t dt \qquad (4.27)$$

This integral can be solved either graphically (see Fig. 4.14) or numerically by taking small slices of p_0; thus, $p_{0,1}, p_{0,2}, \ldots, p_{0,i}, \ldots p_{0,n}$.

$p_{0,i+1} - p_{0,\infty}$	$p_{0,i} - p_{0,\infty}$	$\overline{C_{tot}}$	$\ln\left(\dfrac{p_{0,i+1} - p_{0,\infty}}{p_{0,i} - p_{0,\infty}}\right)$	t
$p_{0,2} - p_{0,\infty}$	$p_{0,1} - p_{0,\infty}$	—	—	—
$p_{0,3} - p_{0,\infty}$	$p_{0,2} - p_{0,\infty}$	—	—	—
$\vdots$	$\vdots$			
$p_{0,n} - p_{0,\infty}$	$p_{0,n-1} - p_{0,\infty}$	—	—	—

To save effort, take a constant ratio; for example, $p = 1, 2, 4, 8, \ldots$

In each interval assume a constant $\overline{C_{tot}}$

Total time $= \Sigma t$

These calculations simplify somewhat

- when there is no leak; thus, when $p_{0,\infty} \to 0$.
- when the pump is located right at the vessel to be evacuated, in which case $C_{tot} \to S_p$.

area $= \dfrac{t}{V}$

$\dfrac{1}{C_{tot}(p_0 - p_{0,\infty})}$

finish   start p_0 Fig. 4.14

Fig. 4.15. Two-stage pumpdown system.

IV. MORE COMPLETE VACUUM SYSTEMS

Often one employs a two-stage pumping system as shown in Fig. 4.15. Let us see how to treat this situation. For section A we write

$$Q_A = C_{\text{tot}, A} p_0 \quad \text{where} \quad \frac{1}{C_{\text{tot}, A}} = \frac{1}{C_{\text{or}, A}} + \frac{1}{C_{\text{line}, A}} + \frac{1}{S_{p, A}}$$

Similarly for section B

$$Q_B = C_{\text{tot}, B} p_3 \quad \text{where} \quad \frac{1}{C_{\text{tot}, B}} = \frac{1}{C_{\text{line}, B}} + \frac{1}{S_{p, B}}$$

Since $Q_A = Q_B$ and $p_3 \gg p_0$ we must make $C_{A, \text{tot}} \gg C_{B, \text{tot}}$. This means that we should design the system so that most of the flow resistance is in pipe B, not A. Therefore

- use a big diameter pipe for line A;
- keep pump A close to the vessel to be evacuated;
- a long, small-diameter tube can be used for line B with but little harm.

V. COMMENTS

This chapter develops the language of molecular flow in the framework of the tank–line–pump system. The field is much broader than this. Here are some additional areas of study:

- The equations for the tank–line–pump situation only holds well if $L/d \geq 100$. For short pipes we may want to modify these expressions. Luckily these are second-order effects.
- Turbulent flow of gases. This situation occurs only very rarely—for high Δp in large pipes.

- Flow in conduits of other shapes: slits, rectangles, annuli, triangles.
- Pumping speed of cold traps.
- Degassing problems—to remove adsorbed gases from metal and glass surfaces.
- Design of more complete vacuum systems.
- High-vacuum gauges and pumps.

REFERENCES AND FURTHER READINGS

A. S. D. Barrett, and B. D. Power, in *Chemical Engineering Practice* (H. W. Cremer and T. Davies, eds.), Vol. 5, Academic Press, New York (1958).

S. Dushman, *Scientific Foundation of Vacuum Technique*, Wiley, New York (1949). Complete and comprehensive.

M. Knudsen, *Ann. Phys.* **28**, 75, 999 (1909). These papers laid the foundations and developed the basics of the whole field.

R. Loevinger, in *Vacuum Equipment and Technique*, (A. Guthrie and R. K. Wakerling, eds.), Chapter 1, McGraw-Hill, New York (1949). A nice, simple treatment of pumping systems.

EXAMPLE 4.1. HIGH-VACUUM FLOW IN A PIPE

Find the speeds S_1, S_2, S_3, and the conductances C_{12}, C_{13}, C_{23} for the pipeline below.

Solution

From the definition of pumping speed we can write, for locations 1, 2, and 3,

$$Q = S_1 p_1 = S_2 p_2 = S_3 p_3$$

Therefore

$$S_1 = \frac{Q}{p_1} = 1\ \mathrm{m^3/s}$$

$$S_2 = \frac{Q}{p_2} = 2\ \mathrm{m^3/s}$$

$$S_3 = \frac{Q}{p_3} = 3\ \mathrm{m^3/s}$$

Next consider the conductance of sections 1–2, 2–3, and 1–3 of the pipe. By definition

$$Q = C_{12}(p_1 - p_2) = C_{23}(p_2 - p_3) = C_{13}(p_1 - p_3)$$

Therefore

$$C_{12} = \frac{Q}{p_1 - p_2} = \frac{12}{12 - 6} = 2\ \mathrm{m^3/s}$$

$$C_{23} = \frac{Q}{p_2 - p_3} = \frac{12}{6 - 4} = 6\ \mathrm{m^3/s}$$

$$C_{13} = \frac{Q}{p_1 - p_3} = \frac{12}{12 - 4} = 1.5\ \mathrm{m^3/s}$$

To check the results:

$$\frac{1}{C_{13}} \overset{?}{=} \frac{1}{C_{12}} + \frac{1}{C_{23}} \quad \text{or} \quad \frac{1}{1.5} \overset{?}{=} \frac{1}{2} + \frac{1}{6} = \frac{2}{3} \quad (\text{correct})$$

EXAMPLE 4.2. CONDITIONS IN A STEADY-STATE VACUUM SYSTEM

A vacuum pump ($S_p = 0.2\ \mathrm{m^3/s}$) is connected by 10 m of 100-mm-i.d. pipe to a large vessel which is to be evacuated of air at room temperature. At a time when the pressure in the tank is 10 mPa
   (a) Calculate the pressure at the mouth of the pump (point 2);
   (b) Determine the pumping speed at the tank (point 0). This is the rate at which air at 10 mPa is being removed from the tank.
   (c) Locate the major resistance to the evacuation.

Solution

We can see from the figure above that $S_2 = S_p$ and $p_2 = p_p$. Then

Including the pump

$$Q = S_0 p_0 = C_{01}(p_0 - p_1) = C_{12}(p_1 - p_2) = S_2 p_2 = C_{tot}(p_0 - 0) \quad (i)$$

— Use underlined terms only —

Imaginary region
beyond the pump

where, from equation (4.22)

$$\frac{1}{C_{tot}} = \frac{1}{C_{or,lam} + C_{or,mol}} + \frac{1}{C_{line,lam} + C_{line,mol}} + \frac{1}{S_2} \quad (ii)$$

From   From   From   From
Eq. (4.11)   Eq. (4.8)   Eq. (4.5)   Eq. (4.6)

Before we proceed to evaluate all the terms let us see whether we are completely in one or the other flow regime. If we are this would simplify matters. At the point of highest pressure, at the tank, we have

$$pd = (0.1)(0.01) = 10^{-3} \text{ Pa m}$$

According to equation (4.1) this condition means that flow is completely in the molecular regime, so we can safely and happily drop the laminar terms in equation (ii). Evaluating the remaining terms gives

$$\frac{1}{C_{tot}} = \frac{1}{91d^2} + \frac{L}{121.3d^3} + \frac{1}{S_2}$$

and on replacing values we find

$$\frac{1}{C_{tot}} = 1.1 + 81.5 + 5 = 87.6 \quad (iii)$$

or

$$C_{tot} = 0.0114 \text{ m}^3/\text{s}$$

Replacing in equation (i) gives

$$p_2 = \frac{C_{tot}\, p_0}{S_2} = \frac{(0.0114)(0.01)}{0.2}$$

$$= 5.7 \times 10^{-4} \text{ Pa} = 0.571 \text{ mPa} \tag{a}$$

The pumping speed at the tank is also given by equation (i). Thus,

$$S_0 = \frac{C_{tot}\, p_0}{p_0} = C_{tot} = 0.0114 \text{ m}^3/\text{s} \tag{b}$$

Equation (iii) shows that the relative resistances are

Entry orifice: $\dfrac{1.1}{87.6} = 1.3\%$

The line: $\dfrac{81.5}{87.6} = 93\%$

The pump: $\dfrac{5}{87.6} = 5.7\%$

Thus, the line provides the major resistance ($\sim 93\%$) $\hspace{2em}$ (c)

NOTE. To speed the evacuation either shorten the pipe or increase the pipe diameter. The latter change is better by far since the pumping speed varies as d^3. Using a bigger pump won't help much. For example, even with the biggest pump in the world, equation (iii) becomes

$$\frac{1}{C_{tot}} = 1.1 + 81.5 + \frac{1}{\infty} = 82.6 \quad \text{vs.} \quad 87.6$$

Thus, the conductance will only increase by about 6%.

If we would have included the laminar pipe resistance term in our calculations our answer would only have changed by about 1%. This justifies our dropping this term.

Since the molecular orifice resistance is only about 1% that of the pipe [see equation (iii)], and the laminar orifice resistance can be expected to be so much smaller than either, the latter can well be ignored.

Actually, to evaluate the laminar conductances of pipe and orifice is awkward and requires a trial-and-error procedure. The next example shows how this is done for pipe flow.

EXAMPLE 4.3. CONDITIONS IN ANOTHER VACUUM SYSTEM

Repeat Example 4.2 with just one change; let the pressure in the tank be 10 Pa.

Solution

At the pipe entry $pd = 1$ Pa m; thus, we are in the regime of laminar flow and should use equation (4.11). However, from Example 4.2 we find that the resistance of the entry is negligible (about 1% of the total) so let us ignore it. Next, following the procedure of Example 4.2 we write

$$Q = S_0 p_0 = S_2 p_2 = C_{tot}(p_0 - 0) \qquad \text{(i)}$$

where

$$\frac{1}{C_{tot}} = \underset{\substack{\text{Neglect}}}{\cancel{\frac{1}{C_{or}}}} + \frac{1}{\dfrac{1364 d^4 \bar{p}}{L} + \dfrac{121.3 d^3}{L}} + \frac{1}{S_2} \qquad \text{(ii)}$$

$$\bar{p} = \frac{p_1 + p_2}{2} \quad \text{where } p_2 \text{ is unknown}$$

Since p_2 is unknown guess that $p_2 = p_1 = 10$ Pa. Then replacing all values in equation (ii) gives

$$\frac{1}{C_{tot}} = \frac{1}{0.1364 + 0.0121} + \frac{1}{0.2} = 6.7326 + 5 = 11.7326$$

or

$$C_{tot} = 0.0852 \text{ m}^3/\text{s}$$

Then from equation (i)

$$p_2 = \frac{C_{tot} p_0}{S_2} = \frac{(0.0852)(10)}{0.2} = 4.2616 \text{ Pa}$$

This value for p_2 does not agree with our guess. So try again. With the help of an astrologer let us guess that p_2 = 3.4 Pa. Then equation (ii) gives

$$\frac{1}{C_{\text{tot}}} = \frac{1}{\dfrac{1364(0.1)^4(10 + 3.4)/2}{10} + \dfrac{121.3(10^{-3})}{10}} + \frac{1}{0.2}$$

$$= \frac{1}{0.0914 + 0.01213} + 5 = 9.66 + 5 = 14.66$$

or

$$C_{\text{tot}} = 0.0682 \text{ m}^3/\text{s}$$

Then equation (4.1) gives

$$p_2 = \frac{C_{\text{tot}} p_0}{S_2} = \frac{(0.0682)(10)}{0.2} = 3.41 \text{ Pa}$$

Our guess was right, so this is the right pressure for p_2, or

$$p_2 = \boxed{3.41 \text{ Pa}} \tag{a}$$

The pumping speed at the tank is also given by equation (i). Thus

$$S_0 = \frac{C_{\text{tot}} p_0}{p_0} = C_{\text{tot}} = \boxed{0.0682 \text{ m}^3/\text{s}} \tag{b}$$

Again the line provides the major resistance, or

$$\frac{9.66}{14.66} \cong \boxed{66\%} \tag{c}$$

NOTE: Laminar flow is the main mechanism causing the movement of fluid in the line. In fact it accounts for

$$\frac{0.0914}{0.0914 + 0.01213} = 88\%$$

of the total flow. Molecular flow accounts for just 12% in the conditions of this problem.

EXAMPLE 4.4. PUMPING DOWN A LEAKY VACUUM SYSTEM

A pump (S_p = 1 m³/s) is placed within a vessel (V = 10 m³) and is pumping it out. However, because of leaks into the vessel, the pressure within the vessel decreases to a limiting value $p_{0,\infty}$ = 1 Pa. Find the leak rate into the vessel.

Solution

When the system reaches steady state

$$Q_{\text{leak}} = Q = C_{\text{tot}} p_0$$

But since there is no line present

$$\frac{1}{C_{\text{tot}}} = \underset{=\,0}{\cancel{\frac{1}{C_{\text{or}}}}} + \underset{=\,0}{\cancel{\frac{1}{C_{\text{line}}}}} + \frac{1}{S_p} = \frac{1}{S_p}.$$

Therefore

$$Q_{\text{leak}} = S_p p_0 = \left(1\ \text{m}^3/\text{s}\right)\left(1\ \text{Pa}\right) = \boxed{1\ \text{Pa m}^3/\text{s}}$$

Thus, 1 m³ of air measured at 1 Pa, or 10 cm³ of air measured at 1 atm, leak into the vessel each second.

PROBLEMS ON VACUUM FLOW

4.1. A molecular still, to be kept at 0.01 Pa, is connected to an oil diffusion pump by a 0.1-m-i.d. line. Data on the pump indicates that it has a pumping speed of 250 lit/s at low pressure. Assume that air at 20°C is being pumped.

(a) What is the speed factor of the pump?

(b) What length of this line can be used and still not reduce the pumping speed at the still to less than 50 lit/s?

4.2. A large vacuum system is connected by 1 m of 0.1-m-i.d. pipe to a 0.4 m³/s vacuum pump. After pumpdown and with the pump working at full speed the pressure in the system is 1 mPa.

(a) What is the pressure at the pump intake?

(b) What is the leak rate of room air into the system? Give this as lit/hr of air at 1 atm.

(c) If the leak rate can be halved what will be the pressure in the system?

4.3. At one end of a tube (10 m long, 0.1-mm-i.d.) the pressure is 1 atm, at the other end the pressure is 1 Pa. The temperature is 20°C.

(a) Find the flow rate of air (measured at 1 atm) through this tube.

(b) Plot the pressure 1/4, 1/2, and 3/4 of the way along the tube.

(c) How will the flow rate change if the tube length is halved?

(d) How will the flow rate change if the tube diameter is doubled?

Ignore the entrance (or orifice) effects.

4.4. An apparatus is connected by 1 m of 4-cm-i.d. glass tubing, free from leaks, to a combined mercury diffusion and mechanical pump (speed = 40 lit/s). Because of small unavoidable leaks in the apparatus itself the minimum pressure attainable in the apparatus is 10 mPa.

(a) What pressure can be maintained in the apparatus if the tubing connecting the pumps to apparatus were shortened to 0.1 m?

(b) What pressure can be maintained in the original apparatus if the pumps were replaced by new ones 10 times as fast?

4.5. During the pumpdown of our vacuum system I noted that it took one day for the pressure to drop from 0.3 Pa to 0.2 Pa. Further pumping reduced the pressure down to 0.1 Pa, but no lower. Yesterday I put extra sealant on what I thought was a leaky joint. Sure enough the pressure in the system started to fall and after 24 hr had gone from 0.1 to 0.06 Pa. Estimate the new limiting low pressure of the system. Since the pump characteristics are not given assume a constant pumping speed at all pressures.

4.6. At present an apparatus is kept at 1 mPa by a vacuum pump and back-up pump running at a speed of 10 l/s, and connected to the apparatus by 1 m of 20-mm-i.d. tubing. We wish to lower the pressure in the apparatus, and three choices come to mind.

(i) Double the tube diameter.

(ii) Shorten the connecting pipe from 1 m to 0.1 m.

(iii) Replace the present pump with a larger one having a pumping speed of 30 lit/s.

Rate these alternatives and present calculations to back your rating.

4.7. A vacuum system consists of a vessel connected by pipe (2–3) to a vacuum pump, more pipe (4–5) to a forepump, and then to the atmosphere as shown.

For the vacuum pump: $S_p = 0.1$ m³/s, $p_3 = 0.2$ Pa, $p_4 = 100$ Pa. Determine the size of forepump which is being used.

4.8. Repeat the previous problem with just one change: the 1 m of 100-mm-i.d. line is to be replaced by 1 m of 1-mm-i.d. line.

4.9. The pumpdown of a 7.5 m³ leak-free vacuum system from 1 Pa to 0.1 Pa takes 2 hr. The pump is connected to the vessel by 1 m of 30-mm-i.d. pipe. Estimate the speed of the pump in this pressure range.

4.10. Our laboratory has a monstrous white elephant, a good-for-nothing three-story-high 12 m³ distillation column. We tried to sell it, to give it away, and finally in desperation we offered $1,000 to anyone who'd remove it. Metal dealers who looked it over agreed that there was a good bit of metal there, but it was too bulky for them.

Well, I think I can help with that. I will have our machinist connect the diffusion pump to the column and evacuate it. From my *Strength of Materials* text I calculate that the column will collapse with a mighty bang into a compact easily moved ball of metal when its interior pressure just reaches 0.1 Pa.

Since our department head is due to make his annual whirlwind visit of our laboratory about 2 pm next Friday wouldn't it be a nice surprise if he were there, maybe even leaning on the tower when it collapsed? What a delicious thought. When should we start evacuating the column so that our honored chief will forever remember his visit?

Data: The pump is connected directly to the tower, and from the manufacturer's data sheet the pump speed is as follows:

Pressure (Pa)	0.8	0.2	0.5	1	2	5	10	20	50	100	1 atm
Pump speed (lit/s)	18	25.5	30.5	32.5	32	19	11	5.5	2.5	1.5	1

One of the key steps in a local company's manufacture of integrated circuits is the low pressure chemical vapor deposition (LPCVD) of exotic materials. This operation takes place in a battery of special reaction chambers, or furnaces, kept at 1 mPa by using an oil diffusion pump followed by a conventional vacuum pump. Without reaction the chamber is capable of maintaining 0.36 mPa.

The operating pressure is reached in a two-step pumpdown:

Step 1. A relatively rapid pumpdown to 100 mPa limited by the conventional vacuum pump. Ignore this time.

Step 2. A longer pumpdown limited by the oil diffusion pump which drops the pressure from 100 mPa to 1 mPa.

Some of the reaction chambers are connected to their pairs of pumps by 1 m of 5-cm line and have a pumpdown time of 42 min; others are connected with 2 m of 5-cm line and have a pumpdown time of 63 min.

Operations are expanding, the maintenance isle is too crowded, and so the staff is thinking of relocating some of the pumps on the next floor. This would require using 8-m lines between chamber and pump pairs.

4.11. How would this affect the pumpdown time?

4.12. What diameter of connecting line should be used to keep the pumpdown time at 42 min? [Problem prepared by Jim McDaniel.]

4.13. A simple way of detecting small leaks in heat exchangers is as follows. Pressurize the unit with air, say to 2 atm absolute, immerse it in hot water containing a sprinkle of detergent to reduce the surface tension, and carefully look for bubbles. This technique is sufficiently sensitive to detect very small leak rates roughly equivalent to forming a 1-mm-diam bubble each minute. How big a hole do you estimate this to represent? Consider the hole to be a pinch point or orifice.

    NOTE: This problem involves material from both Chapters 3 and 4.

At the beginning of the week my bicycle tire (wall thickness = 1.7 mm) contained 1 liter of air at 700 kPa and 20°C. But after 5 days the pressure was down to 690 kPa and I am sure that the air leaked out of just one hole—which was made by *you* when you kicked the tire. Yes, I saw you do it! Estimate the size of the hole

4.14. assuming that the hole is a "pinch point," an orifice.

4.15. assuming that the hole is tubular in form.

    NOTE: These problems involve material from both Chapters 3 and 4.

Chapter 5   NON-NEWTONIAN
FLUIDS

I. CLASSIFICATION OF FLUIDS

Fluids give and flow under stress. Newtonians are the simplest of fluids, and they are characterized by the property that the velocity gradient at a point is proportional to the shear stress at that point, thus

$$(\text{Shear rate}) \, \alpha \, (\text{shear stress}) \quad \text{or} \quad \frac{du}{dy} \, \alpha \tau$$

All other fluids are called non-Newtonians. Air, water, steam, all gases, and most fluids consisting of simple molecules are Newtonians. Dense suspensions, slurries, emulsions, solutions of long-chain polymers, biological fluids,

91

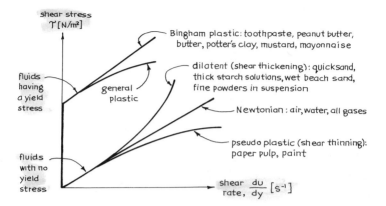

Fig. 5.1. Stress vs. shear rate for various kinds of time-independent non-Newtonians.

liquid foods, paints, clay suspensions, and concrete mix are, as a rule, non-Newtonian.

Non-Newtonians (NNs) can be divided into three broad classes of materials.

1. *Time-independent NNs* for which

$$\begin{pmatrix} \text{Shear} \\ \text{rate} \end{pmatrix} = f\begin{pmatrix} \text{shear stress} \\ \text{alone} \end{pmatrix} \quad \text{or} \quad \frac{du}{dy} = f(\tau \text{ alone})$$

There are a number of types here depending on the form of the τ vs. du/y relationship. These are shown and named in Fig. 5.1.

2. *Time-dependent but nonelastic NNs* are fluids whose present behavior is influenced by what has happened to them in the recent past. For example, tomato catsup which has been resting unmoved for a while will not pour; however, a recently shaken bottle of catsup pours easily. These fluids seem to have a memory which fades with time, thus we can write

$$\begin{pmatrix} \text{Shear} \\ \text{rate} \end{pmatrix} = f\begin{pmatrix} \text{shear} & \text{past history} \\ \text{stress,} & \text{of stress} \end{pmatrix}$$

This behavior is shown in Fig. 5.2.

3. *Viscoelastic NNs* are materials which combine the elastic properties of solids with the flow behavior of fluids, and as examples we have saliva and nearly all biological fluids, concentrated tomato soup, bread dough, and many polymeric solutions. With viscoelastics the τ vs. du/dy diagram only tells part of the story; transient experiments (give the can of tomato soup a quick twist and watch the fluid swish left and right) are needed to characterize their elastic properties.

Fig. 5.2. Stress vs. shear rate for two classes of time-dependent, but nonelastic non-Newtonians.

This chapter develops the flow equations for time independent NNs. For other types of NNs the flow equations, if they can be developed at all, are much more complicated. Fortunately, however, in steady state flow without acceleration (flow in straight pipes without nozzles, bends, orifices, etc.), these fluids can often be treated as time independent, too.

II. SHEAR STRESS AND VISCOSITY

1. For a *Newtonian* the velocity gradient is proportional to the imposed shear stress on the fluid, or

$$\tau = \frac{\mu}{g_c}\left(\frac{du}{dy}\right) \qquad \left[\frac{N}{m^2} = Pa\right] \qquad (5.1)$$

2. For a *Bingham plastic* the shear stress vs. velocity gradient relationship is linear but does not go through the origin, or

$$\tau = \tau_0 + \frac{\eta}{g_c}\left(\frac{du}{dy}\right) \qquad \left[\frac{N}{m^2} = Pa\right] \qquad (5.2)$$

Fig. 5.3. Method of finding the flow parameters of a power law fluid.

Fig. 5.4. Finding the yield stress of a general plastic.

3. *For pseudo plastics and dilatents which follow power law behavior, called power-law fluids,* the relationship between shear stress and velocity gradient is not linear, thus

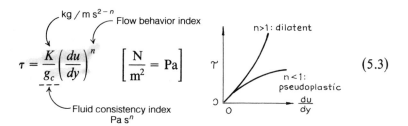

$$\tau = \frac{K}{g_c}\left(\frac{du}{dy}\right)^n \qquad \left[\frac{N}{m^2} = Pa\right] \qquad (5.3)$$

A log–log plot will give K and n, as shown in Fig. 5.3.

There are many other ways of characterizing fluids with no yield stress; however, the power law is a simple representation which reasonably well fits all these fluids.

4. *For general plastics* we have

$$\tau = \tau_0 + \frac{K}{g_c}\left(\frac{du}{dy}\right)^n, \quad \left[\frac{N}{m^2} = Pa\right] \qquad (5.4)$$

The general plastic has characteristics of both Bingham plastics and power-law fluids and it represents a very broad class of fluids, including all previously mentioned time-independent NNs. To determine the three parameters of this type of fluid, τ_0, K, and n, first determine the yield stress τ_0 from Fig. 5.4, then make the log–log plot of Fig. 5.5 to find K and n.

Fig. 5.5. Finding K and n for a general plastic once τ_0 is known.

Fig. 5.6. A better way of finding the yield stress of a general plastic.

Sometimes it is difficult to estimate τ_0 reliably. One useful help is to plot $\sqrt{\tau}$ vs $\sqrt{du/dy}$ for it often gives a straight line near the zero abscissa, as shown in Fig. 5.6. This type of plot was first used by Casson for printer's ink. Thus, the expression of the form

$$\sqrt{\tau} = \sqrt{\tau_0} + K'\sqrt{du/dy} \tag{5.5}$$

is called the Casson equation. It has only a vague association with theory but is useful for finding τ_0.

5. *Comments.* There are many other equation forms for NNs. However, we especially like the above forms since they are easy to use to find f_F and to calculate pumping power requirements. Most other forms are difficult to use.

Time-dependent NNs and viscoelastic NNs have no unique τ vs. du/dy curve. Thus, if we force these fluids into one of the above forms (because it may be too difficult to treat them any other way) we may only safely use the equations for steady-state straight-pipe flow with no acceleration or deceleration of fluid (entrance, exits, orifices, changing pipe size, etc.).

III. FLOW IN PIPES

A. Bingham Plastics

The velocity profile for laminar flow of these materials comes from the shear–velocity relationship

$$\tau = \tau_0 + \frac{\eta}{g_c}\left(\frac{du}{dy}\right) \tag{5.2}$$

and always shows a pluglike core region of diameter

$$d_c = \frac{4\tau_0 L}{\rho \Sigma F} \qquad (5.6)$$

as shown in Fig. 5.7. Proper integration of the velocity profile across the flow tube gives the mean velocity of flow as:

where

$$u = \frac{g_c d^2 \rho \Sigma F}{32\eta L}\left(1 - \frac{4}{3}\mathrm{m} + \frac{1}{3}\mathrm{m}^4\right)$$

Newtonians have this term alone since $\tau_0 = 0$

$$m = \frac{4\tau_0 L}{\rho d \Sigma F} = \frac{\tau_0}{\tau_w}, \quad \text{and} \quad m \le 1 \qquad (5.7)$$

This expression shows that flow is "frozen" when

$$\tau_w < \tau_0 \quad \text{or} \quad \rho\Sigma F < \frac{4\tau_0 L}{d} \quad \text{or} \quad m > 1$$

Looking at it in a different way, when $m = 1$, the term in brackets becomes zero, meaning that the yield stress is nowhere exceeded. Hence, there will be no flow. Thus, the above expression only has meaning when $m < 1$; and when $m \ge 1$, then $u = 0$.

For turbulent flow there is but little information, hence the best we can do today is to use the f_F vs. Re relationship for Newtonians (see Chapter 2).

For Bingham plastics the strategy for using the mechanical energy balance is the same as for Newtonians. Thus, between any two points of a

Fig. 5.7. Velocity profile of Bingham plastics in laminar flow.

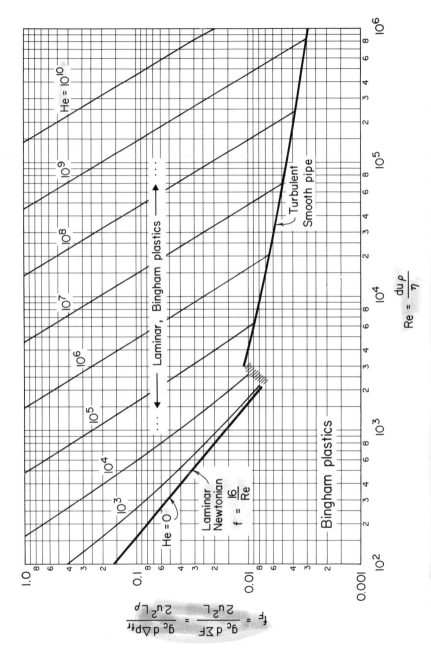

Fig. 5.8. Graph relating frictional loss to flow velocity for Bingham plastics in pipes. [Adapted from B. O. A. Hedstrom, *Ind. Eng. Chem.* **44**, 651 (1952).]

pipe we write equation (1.5) as:

where

$$\frac{g}{g_c}\Delta z + \Delta\left(\frac{u^2}{2g_c}\right) + \int_1^2\frac{dp}{\rho} + W_s + \Sigma F = 0$$

$$\Sigma F = \frac{2f_F Lu^2}{g_c d}$$

and where, with equation (5.7),

$$f_F = f\left[\frac{du\rho}{\eta}, \frac{g_c\tau_0 d^2\rho}{\eta^2}\right] = f(\text{Re}, \text{He})$$

(5.8)

Reynolds number ⟶ Hedstrom number. He = 0 for Newtonians

The relationship between friction factor and Reynolds and Hedstrom numbers is shown in Fig. 5.8. Use it to determine pumping requirements or flow rates in a piece of equipment.

B. Power Law Fluids

For laminar flow the velocity profile, shown in Fig. 5.9, comes from the defining shear expression

$$\tau = \frac{K}{g_c}\left(\frac{du}{dy}\right)^n$$

(5.3)

Integration across the flow tube gives the mean velocity of flow as:

$$u = \frac{g_c d^2\rho\Sigma F}{32KL}\frac{4n}{1+3n}\left(\frac{g_c d\rho\Sigma F}{4KL}\right)^{(1-n)/n}$$

(5.9)

⟵ This term alone for Newtonians

Not much is known about the velocity profile in turbulent flow.

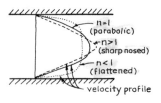

n=1 (parabolic)
n>1 (sharp nosed)
n<1 (flattened)
velocity profile

Fig. 5.9. Velocity profile for power law fluids in laminar flow.

For power-law fluids the mechanical energy balance for flow between any two points of a pipe, equation (1.5), becomes

$$\frac{g}{g_c}\Delta z + \Delta\left(\frac{u^2}{2g_c}\right) + \int_1^2 \frac{dp}{\rho} + W_s + \Sigma F = 0$$

where

$$\Sigma F = \frac{2f_F L u^2}{g_c d}$$

$$f_F = f\left(\begin{array}{c}\text{generalized Reynolds number Re}_{\text{gen}},\\ \text{and probably roughness as well}\end{array}\right)$$

(5.10)

and where, with equation (5.9),

$$\text{Re}_{\text{gen}} = \frac{d^n u^{2-n}\rho}{8^{n-1}K}\left(\frac{4n}{1+3n}\right)^n$$

Figure 5.10 shows the relationship between f_F and Re_{gen} as known today. Note that experiments are not far enough along to include the effect of roughness in this chart.

$$\text{Re}_{\text{gen}} = \frac{d^n u^{2-n}\rho}{8^{n-1}K}\cdot\left(\frac{4n}{3n+1}\right)^n$$

Fig. 5.10. Graph relating frictional loss to flow velocity for power-law fluids in pipes. [Adapted from D. S. Dodge and A. B. Metzner, *AIChE J.* 5, 189 (1959).]

For the flow of power law fluids in circular pipes we use this generalized definition of the Reynolds number because it brings together all the laminar flow lines on the f_F vs. Re chart. It should be noted however that this definition is not useful for flow in other shapes of flow channel or for other than power law fluids.

C. General Plastics

The velocity profile is complex and comes from the shear–velocity relationship:

$$\tau = \tau_0 + \frac{K}{g_c}\left(\frac{du}{dy}\right)^n \qquad (5.4)$$

A plug flow core section is always present here as shown in Fig. 5.11.

For laminar flow integration of the velocity profile expression gives the mean velocity of flow as

$$u = \underbrace{\frac{g_c d^2\rho\Sigma F}{32KL}}_{\substack{u = 0 \text{ when} \\ m \geq 1}} \underbrace{\left(\frac{g_c d\rho\Sigma F}{4KL}\right)^{(1-n)/n}}_{\text{This term alone for Newtonians}} 4n(1 - m)^{(n+1)/n}$$

$$\times\left[\frac{(1 - m)^2}{1 + 3n} + \frac{2m(1 - m)}{1 + 2n} + \frac{m^2}{1 + n}\right] \qquad (5.11)$$

where

$$m = \frac{4\tau_0 L}{\rho d\Sigma F} = \frac{\tau_0}{\tau_{\text{wall}}} \quad \text{and} \quad m \leq 1$$

This laminar flow expression reduces to the corresponding expressions for Bingham plastics (when $n = 1$), for power law fluids (when $\tau_0 = 0$), and for Newtonians (when both $n = 1$ and $\tau_0 = 0$).

The f_F vs. Re chart for general plastics has no yet been developed. So to evaluate the frictional loss in flow we must use equation (5.11) for

Fig. 5.11. Velocity profile for general plastics.

laminar flow, and we must somehow interpolate between Bingham plastics and power law fluids for turbulent flow.

D. Comments on Flow in Pipes

Many NNs are rather viscous liquids whose flows are in the laminar regime. Consequently the flow equations for this regime are of particular interest.

The frictional pressure drop Δp_{fr} is sometimes used in place of the frictional loss ΣF. These terms are related by the mechanical energy balance of Chapter 1. For liquids where $\rho \cong$ const

$$\Delta p_{\mathrm{fr}} = \rho \Sigma F = -\left[\frac{g\rho \Delta z}{g_c} + \Delta\left(\frac{\rho u^2}{2g_c} \right) + \Delta p \right] \qquad (5.12)$$

The frictional loss is always positive

$(p_2 - p_1)_{\text{measured}}$

In the often encountered special case where the kinetic and potential energy terms can also be ignored we have

$$\Delta p_{\mathrm{fr}} = \rho \Sigma F = -\Delta p_{\text{measured}} \qquad (5.13)$$

IV. DETERMINING FLOW PROPERTIES OF FLUIDS

Three classes of devices are commonly used to measure the flow properties of non-Newtonians. These are shown in Figs. 5.12, 5.13, and 5.14.

Fig. 5.12. Tube viscometer or extrusion rheometer: the simplest of devices to measure the flow properties of non-Newtonians.

Fig. 5.13. Concentric cylinder rotary viscometer: especially useful are the narrow gap viscometer where $r_1/r_2 \to 1$, and the infinite medium viscometer where $r_2 \to \infty$.

Fig. 5.14. Cone and plate viscometer.   Fig. 5.15. Narrow gap viscometer.

Let us see how to evaluate the flow properties of a fluid from experiments made in two variations of the concentric cylinder viscometer and in the tube viscometer.

A. Narrow Gap Viscometer

With $r_1/r_2 \rightarrow 1$, as shown in Fig. 5.15, we can safely use the flat-plate approximation, in which case the torque is related to the shear by

$$\text{(Torque)} = \left(\begin{array}{c}\text{shear} \\ \text{stress}\end{array}\right)\left(\begin{array}{c}\text{radial} \\ \text{distance} \\ \text{of gap}\end{array}\right)\left(\begin{array}{c}\text{wetted} \\ \text{area}\end{array}\right) = \tau \cdot r \cdot 2\pi r L, \quad [\text{J}] \quad (5.14)$$

while the velocity gradient is given by

$$\frac{du}{dr} = \frac{2\pi r N}{\Delta r} \quad [\text{s}^{-1}] \tag{5.15}$$

All we need do is measure the torque at different rotation rates. This gives a series of τ vs. du/dr measurements from which we can find the flow parameters directly and simply.

Remarks. (a) This device has severe frictional effects which heat the fluid, changing its properties. Thus cooling is often necessary.

(b) End effects such as friction at the bottom of the bob are accounted for by measuring the torque for a Newtonian of known viscosity and then finding a value for effective length L_e such that the following shear–torque equation is satisfied

$$\text{(Torque)} = 2\pi r^2 L_e \frac{\mu}{g_c}\left(-\frac{du}{dr}\right) \quad [\text{N m}] \tag{5.16}$$

(c) This device is not too flexible in its ability to treat a large range in shear stress.

(d) For other minor corrections see Skelland (1967).

Fig. 5.16. Infinite medium viscometer.

B. Cylinder in an Infinite Medium

The shear–torque relationship for this device, shown in Fig. 5.16, is

$$\tau = \frac{\text{torque}}{2\pi r^2 L} \qquad (5.17)$$

while the velocity gradient at the surface of the spinning cylinder is calculated to be

$$\frac{du}{dr} = 4\pi N \left[\frac{d(\log N)}{d(\log \text{torque})} \right] \qquad (5.18)$$

Experimentally all we need do is measure the torque at a number of N values, plot on a log–log scale, and evaluate the slope at various N values. With equations (5.17) and (5.18) this gives the τ vs. du/dr curve directly. Thus, this device can be used to find the flow parameters for all time-independent NNs.

Remarks. (a) This device is quick and easy to use.

(b) To correct for end effects use an equivalent length L_e, as with the narrow gap viscometer.

(c) To insure that the outer cylinder is large enough for the infinite medium approximation to apply, use two different sizes of outer cylinders and compare results.

(d) When the gap is somewhere between the above two extremes the equations become complicated.

(e) For additional minor corrections see Skelland (1967).

C. Tube Viscometer

For laminar flow a force balance on a section of tube of length L gives

$$\tau_w = \frac{d\Delta p_{\text{fr}}}{4L} = \frac{d\rho \Sigma F}{4L} \qquad (5.19)$$

It has also been shown by Skelland (1967) that at the wall of the tube

$$\left(-\frac{du}{dr}\right)_w = \frac{8u}{d}\left(\frac{3n'+1}{4n'}\right) \tag{5.20}$$

where

$$n' = \frac{d\log(d\rho\Sigma F/4L)}{d\log(8u/d)} \tag{5.21}$$

The procedure for finding the flow characteristics of a power-law fluid is then as follows:

1. Make a series of runs using different tube diameter, tube length, or pressure drop. For each run measure Δp and u, then evaluate both $d\rho\Sigma F/4L$ and $8u/d$.

2. Make a log–log plot of $d\rho\Sigma F/4L$ vs. $8u/d$ and determine its slope n' and intercept K'/g_c, as shown in Fig. 5.17.

- If $n' = 1$ then the fluid is a Newtonian with viscosity $\mu = K'$.
- If n' is constant, but different from unity, then the fluid is a power law fluid with flow parameters

$$n = n'$$

$$K = K'\left(\frac{4n}{3n+1}\right)^n$$

- If n' is not constant calculate τ_w and $(-du/dr)_w$ for each point from equations (5.19), (5.20), and (5.21), and construct the τ_w vs. $(-du/dr_w)$ chart. From this find the type of fluid at hand and evaluate its flow parameters.

Remarks. (a) This device is very flexible in that it can explore a wide range of shear values. For slurries one should have an agitator in the reservoir.

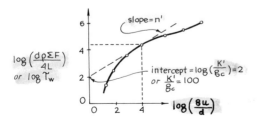

Fig. 5.17. Evaluation of flow parameters at a given flow rate in a pipe.

(b) The above equations only apply to the laminar flow regime. Be sure to verify that flow is in this regime by evaluating the generalized Reynolds number

$$
\text{Re}_{\text{gen}} = \frac{d^n u^{2-n} \rho}{8^{n-1} K} \left(\frac{4n}{1 + 3n} \right)^n
$$

for the run at the highest flow rate, after having found the flow parameters.
(c) In the above expressions the term $\rho \Sigma F$ is evaluated with equations (5.12) and (5.13). To account for the kinetic energy and potential energy effects in these expressions, plus slip effects, once again see Skelland (1967).

V. DISCUSSION ON NON-NEWTONIANS

A. Materials Having a Yield Stress, Such as Bingham Plastics

These materials show both solidlike (elastic) and liquidlike (plastic) behavior depending on the intensity of shear acting on them. Many substances, normally treated as solids exhibit this elastic–plastic behavior; metals, for example. With sufficient shear they give, they flow, and they can be extruded, shaped, punched, etc. On the other hand glass and many rocks do not have a yield stress on man's time scale. But even they do on the geological scale.

Table 5.1. Flow Parameters of Some Familiar Bingham Plastics

Material	Yield stress, τ_0 (Pa = N/m²)	Plastic viscosity, η (kg/m s)
Catsup (30°C)	14	0.08
Mustard (30°C)	38	0.25
Oleomargarine (30°C)	51	0.72
Mayonnaise (30°C)	85	0.63
Butter, on a warm day, very soft and nearly melted	10–20	
Butter, just out of refrigerator, hard, but just spreadable	100–150	
Lead (20°C)	1.3×10^7	
Solder (20°C)	2.7×10^7	
Copper (20°C)	7×10^7	
Various iron and steel (20°C)	$20–50 \times 10^7$	
Titanium alloy (20°C)	120×10^7	

On the microscopic level, this elastic–plastic behavior can occur either in *metals* by atomic displacement, or the sliding of molecules over each other, or in *two-phase systems*, one finely dispersed in the other with large surface forces acting between phases. Examples of such two-phase systems are

- Mayonnaise—oil dispersed in aqueous solution.
- Whipped cream—air dispersed in protein solution.
- Margarine—fat crystals dispersed in oil.
- Chocolate—sugar and cocoa dispersed in cocoa butter (35%).
- Mashed potatoes—water (90%) dispersed in food material.
- Toothpaste—chalk dispersed in water.

The Bingham plastic is the simplest representation for materials having a yield stress, and Table 5.1 gives values of the flow parameters for some commonly met materials of this type.

B. Power-Law Fluids

The power law is a convenient representation for many fluids, and Table 5.2 shows values of the flow parameters for some familiar substances which can reasonably be represented by this model.

C. Thoughts on the Classification of Materials

1. The observation that the measured viscosity of a fluid changes with shear rate or velocity is a sure sign of non-Newtonian behavior (see Figs. 5.1 and 5.2).

Table 5.2. Flow Parameters of Some Familiar Power-Law Fluids

Material	n $(-)$	K $(kg/m\ s^{2-n})$
Applesauce, different recipes (24°C)	0.41	0.66
(24°C)	0.65	0.50
Banana puree, different samples (24°C)	0.46	6.5
(24°C)	0.33	10.7
Human blood	0.89	0.00384
Soups and sauces	0.51	3.6–5.6
Tomato juice (5.8% solids, 32°C)	0.59	0.22
(30% solids, 32°C)	0.40	18.7
4% paper pulp in water (20–30 cm pipe)	0.575	20.7
33% lime in water (2–5 cm pipe)	0.171	7.16
15% carboxy-methyl-cellulose in water (2–4 cm pipe)	0.554	3.13

2. Power-law fluids and Bingham plastics are the simplest models for non-Newtonians. Many more complicated ones are available. Luckily these two simple approximations are often quite satisfactory for engineering purposes. Even time-dependent and other more complex materials flowing at steady state in pipes can often be treated by these simple models.

3. Sometimes the rate of displacement, du/dy, will determine whether a material behaves as fluid or solid. High rates will cause the material to break, while low rates will result in flow. "Silly Putty" and concentrated starch–cold water mixtures are familiar examples. Even liquid water breaks at high shear. On the other hand, even glass will give and flow at room temperature if we wait long enough. For example, window panes from medieval times are thinner at the top than at the bottom.

4. Most biologic fluids are NNs and must be treated as such.

5. Most NNs can be classified in more than one way depending on how they are being processed.

6. We have only presented a simple approach to NNs. The whole question of time dependency (shake vigorously and catsup thins, let it stand and it thickens) and viscoelastic behavior is something we do not touch.

REFERENCES AND RELATED READINGS

A. H. P. Skelland, *Non-Newtonian Flow and Heat Transfer*, Wiley, New York (1967). Excellent primary reference; go here first.

S. E. Charm, *The Fundamentals of Food Engineering*, Second ed., Ch. 3, Avi, Westport (1971). Hurriedly written and difficult to follow but has much useful material.

H. G. Muller, *An Introduction to Food Rheology*, Crane Russak, New York (1973). Delightful gem, even recommended for bedtime reading.

G. W. Govier and K. Aziz, *The Flow of Complex Mixtures in Pipes*, Von Nostrand Reinhold, New York (1972). Afraid to leave anything out, has everything; an omni-book.

R. B. Bird, G. C. Dai, and B. J. Yarusso, *Reviews in Chemical Engineering*, **1**, 1 (1982). A good, well-organized review of theory and experiment on flow and heat transfer of non-Newtonians having a yield stress.

NOTE: Most of the data in this chapter on the flow properties of non-Newtonians are gathered from Skelland, Charm, and Muller and from the *Handbook of Chemistry and Physics*, 47th ed., Chemical Rubber, Boca Raton, FL (1966).

EXAMPLE 5.1. FLOW OF A BINGHAM PLASTIC FROM A TANK

A Bingham plastic ($\tau_0 = 20$ Pa, $\eta = 0.02$ kg/m s, $\rho = 2000$ kg/m^3) discharges from the bottom of a storage tank through a horizontal 100-mm-i.d. pipe of equivalent length of 19.6 m. What head of fluid h will give an outflow velocity of 1 m/s?

Solution

We can find the head either by using the design charts for Bingham plastics (Fig. 5.8) or by using the flow equation (5.7). Let us find the head both ways.

Method A, using the design chart of Fig. 5.8. For this determine

$$He = \frac{g_c \tau_0 d^2 \rho}{\eta^2} = \frac{(1)(20)(0.1)^2(2000)}{(0.02)^2} = 10^6$$

$$Re = \frac{du\rho}{\eta} = \frac{(0.1)(1)(2000)}{0.02} = 10^4$$

$$f_F = 0.025 \quad \text{(from Fig. 8)}$$

To find the head needed apply the mechanical energy balance between points 1 and 2 to give

$$\Delta z + \underset{\approx\,0}{\cancel{\frac{\Delta u^2}{2g}}} + \underset{=\,0}{\cancel{\frac{g_c}{g}\frac{\Delta p}{\rho}}} + \underset{=\,0}{\cancel{\frac{g}{g}W_s}} + \frac{g_c}{g}\Sigma F = 0$$

or

$$h = z_1 - z_2 = \frac{2f_F u^2 L}{gd}$$

$$= \frac{2(0.025)(1)^2(19.6)}{0.1(9.8)} = \boxed{1\ \text{m}}$$

NOTE: In this solution we ignored the kinetic energy term because it was felt that 1 m/s represented a minor energy effect. If this term were included we'd have

$$h = \frac{2f_F L u_2^2}{gd} + \frac{u_2^2}{2g} = 1 + \frac{1^2}{2(9.8)} = 1.05 \text{ m}$$

or a 5% correction.

Method B, using the flow equation (5.7). First we need to evaluate a number of terms. Between points 1 and 2 the mechanical energy balance gives

$$\Sigma F = -\left[\frac{g \Delta z}{g_c} + \cancel{\frac{\Delta u^2}{2 g_c}}^{= 0} + \cancel{\frac{\Delta p}{\rho}}^{= 0} + \cancel{W_s}^{= 0} \right]$$

$$= -\frac{9.8(\Delta z)}{1} = 9.8h \frac{\text{J}}{\text{kg}} \qquad \Sigma F = \frac{4\tau_0 L}{\rho d\, m}$$

then

$$m = \frac{4\tau_0 L}{\rho\, d \Sigma F} = \frac{4(20)(19.6)}{(2000)(0.1)9.8h} = \frac{0.8}{h}$$

Replacing all the known values in equation (5.7) gives

$$1 = \frac{(1)(0.1)^2(2000)9.8h}{128(0.02)(19.6)} \left[1 - \frac{4}{3}\frac{0.8}{h} + \frac{1}{3}\frac{(0.8)^4}{h^4} \right]$$

or

$$h^4 - 1.1306 h^3 + 0.13653 = 0$$

or

$$h = 0.99 \text{ m}$$

which is close to the answer in method A. Actually the answer to method A is a bit off because of chart reading error.

EXAMPLE 5.2. TRANSPORTING COAL BY PIPELINE

In northern Arizona the Peabody Coal Co. transports coal ($\rho = 1500$ kg/m^3) by slurry pipeline. For this purpose coal is crushed and pulverized to minus 8 mesh and is then pumped 440 km as a 50%-by-weight slurry in a 0.45-m-i.d. pipeline having four pumping stations. Transit time is 3 days.

(a) What pumping power is needed if the pump and motor are 70% efficient?

(b) What is the pumping cost/ton to transport the coal from the mine to destination?

Data:

- A 50% by weight coal slurry behaves as a power-law fluid with $n = 0.2$ and $K = 0.58$, in SI units.
- Electricity costs 3¢/kW hr.

[See *Chemical and Engineering News*, p. 17 (April 15, 1974) for more information on this operation.]

Solution

From the data we have

$$u = \frac{400 \text{ km}}{3 \text{ days}} \left(\frac{1 \text{ day}}{24 \times 3600 \text{ s}} \right) \left(\frac{1000 \text{ m}}{1 \text{ km}} \right) = 1.7 \frac{\text{m}}{\text{s}}$$

Mass flow rate of slurry transferred:

$$\dot{m} = uA\rho = (1.7) \left[\frac{\pi}{4} (0.45)^2 \right] (1200) = 324 \frac{\text{kg}}{\text{s}}$$

Density of a 50% by weight mixture, after much fiddling about, is found to be

$$\bar{\rho} = \frac{2\rho_1\rho_2}{\rho_1 + \rho_2} = \frac{2(1000)(1500)}{1000 + 1500} = 1200 \frac{\text{kg}}{\text{m}^3}$$

Since the slurry is a power law fluid

$$\text{Re}_{\text{gen}} = \frac{d^n u^{2-n} \rho}{8^{n-1} K} \left(\frac{4n}{1 + 3n} \right)^n$$

$$= \frac{(0.45)^{0.2} (1.7)^{1.8} (1200)}{8^{-0.8} (0.58)} \left[\frac{4(0.2)}{1 + 3(0.2)} \right]^{0.2} = 21,060$$

So from Fig. 5.10 we find

$$f_F = 0.002$$

Now to the mechanical energy balance. Across the whole pipeline (point 1 to 3) we may write

$$\underbrace{\frac{g\Delta z}{g_c}}_{=0} + \underbrace{\frac{\Delta u^2}{2g_c}}_{=0} + \underbrace{\frac{\Delta p}{\rho}}_{=0} + W_s + \Sigma F = 0$$

Let us explain the cancelled terms. In the third term p_1 and p_3 are both at 1 atm so $\Delta p = 0$. Next, because of the great length of the pipeline, the frictional losses therein should dominate and overwhelm the other losses. Thus, the kinetic energy losses should be negligible, and entrance effects and elevation of the storage tank can safely be ignored. Finally, the difference in elevation, since it is not given, will be ignored. This leaves us with

$$-W_s = \Sigma F = \frac{2f_F u^2 L}{g_c d}$$

$$= \frac{2(0.002)(1.7)^2(440,000)}{(1)(0.45)} = 11,303 \frac{J}{kg}$$

and the actual total power requirement

$$-\dot{W}_s = (11.3 \text{ kJ/kg})(324 \text{ kg/s})(1/0.7) = \boxed{5239 \text{ kW}}$$

or

$$-\dot{W}_s = \boxed{1310 \text{ kW/pumping station}}$$

$$\left.\begin{array}{c} \\ \\ \end{array}\right\} \quad \text{(a)}$$

The cost of transporting coal by pipeline is then

$$(5239 \text{ kW})\left(0.03 \frac{\$}{\text{kW hr}}\right)\left(\frac{\text{hr}}{3600 \text{ s}}\right)\frac{1 \text{ s}}{(324/2) \text{ kg coal}}\left(1000 \frac{\text{kg}}{\text{ton}}\right)$$

$$= \boxed{27\text{¢/ton of coal}} \quad \text{(b)}$$

PROBLEMS ON NON-NEWTONIANS

5.1. Consider a Bingham plastic flowing in a horizontal pipe. If the pressure drop from end to end of the pipe is lowered, then flow slows naturally. Eventually, if Δp is lowered far enough, a critical point is reached where flow stops and

material "freezes" in the pipe. Determine this critical Δp for tomato catsup
flowing in a horizontal pipe ($L = 10$ m, $d = 10$ cm).

5.2. If flow of a Bingham plastic just "freezes" in a 10 m length of horizontal
10-cm-i.d. pipe, what length of 20-cm-i.d. pipe would cause the fluid to just
freeze for the same overall pressure drop across the pipe?

5.3. What diameter of vertical tube would allow mayonnaise ($\rho = 1200$ kg/m^3) to
flow under its own weight?

5.4. A 3-mm-i.d. tube 100 mm long is connected to the bottom of a vat of mustard
and is pointing straight down. When the vat is full (depth of 1 m) mustard
flows out the tube, but when the depth in the tank falls to 0.4 m flow stops.
From the above information find the yield stress of mustard, a Bingham
plastic of density $\rho = 1200$ kg/m^3.

A Bingham plastic ($\tau_0 = 20$ Pa, $\eta = 0.2$ kg/m s, $\rho = 2000$ kg/m^3) discharges from
the bottom of a storage tank through a horizontal 0.1-m-i.d. pipe. Determine the
outflow velocity from the pipe if the pipe is located 10 m below the liquid level in
the tank and has an equivalent length of

5.5. 4.9 m.

5.6. 19.6 m.

5.7. Leer toothpaste is to be pumped through a 50-mm-i.d. stainless-steel tube
from the ingredients-blending machine to the toothpaste-tube-filling machine.
The equivalent length of line, including bends, fittings, and entrance and exit
losses, is 10 m and the mean velocity of flow is to be 1 m/s.
(a) What pressure difference (in atm) will give this flow rate?
(b) What size motor will do the job for a pump–motor efficiency of 30%?
Data: Leer can be considered to be a Bingham plastic with the following
properties: $\rho = 1600$ kg/m^3; $\tau_0 = 200$ Pa; $\eta = 10$ kg/m s.

5.8. We wish to pump homogenized soy bean butter ($\rho = 1250$ kg/m^3; $\tau_0 = 80$ Pa; $\eta = 1$ kg/m s) from a storage tank on the upper floor of our little
co-op factory to the packaging department below. What size of pump and
motor, at 50% efficiency, should we build into the line to guarantee that the
flow rate never gets below 0.8 m/s. See drawing below for additional data.

5.9. At the time of the year when honey is harvested we plan to use the equipment of Problem 5.8 to pump spiced, blended honey at a velocity never lower than 0.8 m/s from the mixing vat to the packaging department. The pump and motor are expected to be 50% efficient for this operation. What size of pump and motor should we order?

Data: Flavored honey is a power-law fluid with: $n = 2$; $K = 5/98$ kg/m; $\rho = 1250$ kg/m^3.

Paint is to be pumped at 1 m/s through a horizontal 1-cm-i.d. pipe 25 m long. Find the size of pump–motor, at 40% efficiency, needed.

5.10. Solve using the design charts given in this chapter.
5.11. Solve using the flow equation given in this chapter.

Data: This paint follows power-law fluid behavior with $n = 0.5$, $K = 2.53$ kg/m s$^{3/2}$, and $\rho = 2000$ kg/m^3.

5.12. Yummy Oriental Delicacies, Inc., has completely changed American home eating habits, and of the 23 Chinese dishes that it prepares and markets, sweet and sour pork dinners are the overwhelming favorite. Grocery stores just can't meet the demand and a black market is developing for this item. Yummy will have to speed production of this item as soon as possible, and management has decided on 10 times the present production rate.

Yummy's production line for sweet and sour pork, simplified somewhat, is shown above. Your job is to replace delivery tube A with a larger size of tube. The company's astrologer–acupuncturist says that doubling the tube diameter (if done between the 10th and 15th of the month) will do the job. The chef is not quite sure and asks you to verify this recommendation. Specifically, if you double the tube diameter what will happen to the delivery rate of the thick starch solution?

Data: As a first approximation treat the starch solution as a power-law dilatent with $n = 2$.

5.13. A 4% paper pulp slurry is to be pumped from a well-mixed storage tank through a 20-cm-i.d. pipe to a processing tank. The liquid level in the processing tank is 10 m higher than in the storage tank, and the equivalent length of connecting pipe is 40 m. The pump and motor on the line are rated at 25 kW and are 50% efficient overall. With this arrangement, as shown below, calculate the expected flow rate of slurry in m^3/s.

5.14. The flow rate of the previous problem is too low. We wish to raise it to 0.2 m^3/s. What size of pump and motor (at 50% efficiency) will do the job?

5.15. *Banana pipeline.* Why ship bananas by boat from Honduras to the U.S.? Look at all the hauling, loading onto ships, squashing, spoilage, worry about rats and cockroaches, unloading, trucking, handling, labor, etc. Why not peel the bananas in Honduras and pump them as a puree by pipeline straight to Chicago, just 5000 km away, then add a touch of vegetable glue to reform the bananas into any desired shape. What an exciting opportunity for creative banana reshaping. However, before plunging into banana reformation, let us see if this proposal is economical. So ignoring capital costs in building the pipeline, estimate the pumping cost for such an operation. Give costs as $/yr and ¢/kg of shipped bananas.

　　Data: Assume: Pump and motor are 50% efficient overall
　　　　　　　　Electricity costs 3.6¢/kW hr, or 1¢/MJ
　　　　　　　　Pipe size = 10-cm-i.d.
　　　　　　　　Mean banana velocity = 1 m/s
　　Flow characteristic of banana puree is given as

$$\tau = 6.3(du/dy)^{1/3} \text{ (all in SI units.)}$$

5.16. Chicago and the Midwest have really taken to our new reformed bananas (see above problem) so we contemplate installing more pumping stations and raising the flow rate in our pipeline to 6 m/s. With this change what does it cost per kilogram to pump the bananas from Honduras to Chicago?

5.17. Consider the following facts.

　　• West Asian countries presently ship vast quantities of oil all over the world in giant tankers. These return empty—what a waste.

　　• These countries also have lots of natural gas which they don't know what to do with.

- Some of the oil importing countries have much iron ore. The fines are not much liked because they cannot be used directly in blast furnaces because of clogging.
- In recent years a number of firms have developed processes for reducing pelletized iron ore fines with natural gas.

An obvious thought—why not fill the empty returning tankers with a thick iron ore slurry and make steel right there in West Asia. It would at one fell swoop solve the problems of empty tankers, wasted ore fines, wasted natural gas, and at the same time would produce steel for these growing economies.

Let us explore one tiny aspect of the overall process, the pumping of the dense slurry of iron ore into the hold of the tankers from buried storage tanks vented to the atmosphere. What size of pump and motor is needed (at 33% efficiency) for a flow rate of 2 m/sec of dense slurry in the 0.3 m pipe.

Data: Density of slurry: $\rho = 3000$ kg/m^3. The slurry is definitely a non-Newtonian, represented as a power-law fluid with $K = 3$ kg/m s^{2-n} and $n = 0.15$.

5.18. Dutch Masters paint pigment is made of TiO_2 powder in water plus dispersant (to keep the solid from settling), thickener (a cellulose derivative), formaldehyde (to keep bugs from eating the thickener), plus this and that. The final result, after much difficult mixing and stirring, is a beautiful white Bingham pseudoplastic whose flow characteristics are given by

$$\tau = 20 + 2(du/dy)^{0.5}$$

The pigment ($\rho = 1700$ kg/m^3) flows from vat to tank car through 25 m of 20-cm pipe. How long would it take to fill a tank car (36 m^3) by gravity alone if the level of pigment in the vat is about 6 m above the pipe exit?

5.19. *Coal to Texas.* Texas Eastern Corp. is planning to pump coal ($\rho = 1500$ kg/m^3) in a 0.96-m-i.d. slurry pipeline from the coal mining regions of Montana (elevation = 1400 m) to the Texas Gulf coast (elevation = 30 m), a distance of 3000 km. Water from the Little Bighorn River (25×10^6 m^3/yr will be used to make the 50% by volume slurry, which is a power-law fluid ($n = 0.2$; $K = 0.65$ kg/m s$^{1.8}$).

What will be the cost per ton of coal transported this way if power costs 3¢/kW hr and if the whole pumping system is 50% efficient? [Information from *Chem. Eng. News*, p. 20 (March 12, 1979).]

5.20. We plan to produce and market a brilliant blindingly bright new toothpaste called Leer. A small pilot plant is already built and samples of Leer are available for testing. In the full-size plant we will have to pump Leer around, and to do this effectively we need to know its flow properties. For this we introduce Leer into a rotating cup viscometer of dimensions shown here:

torque indicator

put Leer heer

10cm

rotate outer cup

9.9 cm

10.1 cm

We find that the cup is able to rotate only when the torque exceeds $\pi/10$ N m; and the cup rotates at 3.8 rpm when the torque is $\pi/5$ N m. What kind of fluid is Leer and what are the values of its flow parameters?

5.21. Find the flow properties of a scrumptious 5-ton batch of warm chocolate, after 72 hr of blending, from the following data obtained in a narrow gap viscometer ($r_1 = 25$ mm; $r_2 = 28$ mm; $L_e = 76.4$ mm)

Torque (N m)	0.0051	0.0077	0.0158	0.0414
Rotational rate (min^{-1})	Just begins to rotate	0.39	2.62	14.81

[Data inspired by Charm, p. 63 (1971).]

5.22. A cylinder ($r = 0.95$ cm, $L_e = 4$ cm) is lowered into a vat of concentrated orange juice at 0°C, rotated, and the torque is measured, with the following results

Rotational rate (sec^{-1})	0.1	0.2	0.5	1.0
Torque (N m)	42×10^{-6}	63×10^{-6}	107×10^{-6}	152×10^{-6}

Find the flow characteristics of this sample of orange juice. [Data from Charm, p. 64 (1971).]

5.23. Find the flow characteristics of pureed missionary soup (Niugini variety) from the following data taken in a tube viscometer.

d (mm)	L (cm)	Δp (MPa)	$\dot{v}$ (cm^3/s)
0.8	10	1.6	0.05
0.8	10	5	0.5
8	200	1	5

[Data from Port Moresby.]

The flow of a suspected power-law fluid ($\rho = 1000 \text{ kg/m}^2$) is being investigated in a capillary tube viscometer (tube diameter $= 1$ mm; tube length $= 100$ mm). Two runs are made with results shown below.

(a) Find the flow parameters of this fluid.

(b) What would you name this fluid?

(c) The equations for this viscometer are only applicable to laminar flow, so calculate the Reynolds number for the run at the higher flow rate to verify that this condition is satisfied.

		Flow rate	Pressure drop across the tube
5.24.	Run 1	3.535 kg/hr	4 MPa
	Run 2	0.03535 kg/hr	0.8 MPa
5.25.	Run 1	0.3535 kg/hr	4 MPa
	Run 2	0.03535 kg/hr	0.8 MPa

5.26. Find the flow characteristics of tomato paste from the following data, taken in a tube viscometer: $L = 1.22$ m; $d = 12.7$ mm; height of tomato paste in the reservoir $= 0.11$ m and $\rho = 1120 \text{ kg/m}^3$.

$\dot{v}$ (cm^3/s)	0.1	0.5	1.3	4.3
$-\Delta p$ (Pa)	19600	27500	34800	43800

[This data comes from Charm, p. 62 (1971).]

5.27. Observation shows that tomato paste has a yield stress, so develop an equation which includes this factor to represent the flow data of the previous problem.

5.28. The strained and cooled (10°C) juice of Rocky Mountain oyster stew is reported to have flow properties displayed below. How would you describe the viscous properties of this regional delicacy? Give an equation if you can.

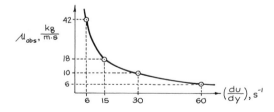

5.29. We have already determined the flow properties of Leer toothpaste (Bingham plastic: $\tau_0 = 200 \text{ N/m}^2$; $\eta = 10 \text{ kg/m s}$), and we have pumped it from the blender to the toothpaste-tube-filling machine. Now consider the design of the toothpaste tube itself.

For a reasonable ribbon of toothpaste it is well accepted that the nozzle of the tube should be 7-mm-i.d. Also, the "finger-force" needed to push the toothpaste out of the tube should not be too large or too small. How long should the tube nozzle be so that toothpaste will just squeeze out when the pressure on the toothpaste is 2 kPa?

Chapter 6   FLOW THROUGH PACKED BEDS

There are two quite different types of porous media:

- *Packed beds*: These include rock piles, sand filters, soil in flower pots, cigarettes, and absorption columns. For good gas–liquid contacting, absorption columns are usually packed with specially shaped ceramic, plastic, or metal objects such as rings and saddles which have both a large surface area and high voidage; thus, they have a low resistance to flow.
- *Porous solids*: These include both natural porous materials such as oil-bearing rock and pumice, and prepared structures such as fused alumina particles, polyurethane foam sponges, foam rubber mattresses, etc.

This chapter concentrates on flow in packed beds.

I. CHARACTERIZATION OF A PACKED BED

A. Sphericity ϕ of a Particle

The sphericity is the most useful single measure for characterizing the shape of irregular and other nonspherical particles. It is defined as

$$\phi = \left(\frac{\text{surface of sphere}}{\text{surface of particle}} \right)_{\substack{\text{same} \\ \text{volume}}} = \frac{\bigcirc}{\text{▱}} \leq 1 \qquad (6.1)$$

Table 6.1 shows values of sphericity for various familiar shaped particles.

B. Particle Size, d_p

When we measure the size of a spherical particle, with a ruler, by screen analysis, or whatever, we know what the measurement means. But with nonspherical particles we have difficulties. When we come up with a size for the particle of Fig. 6.1 do we mean dimension "a" or "b" or something else?

Table 6.1. Sphericity of Particles[a]

Particle shape	Sphericity ϕ
Sphere	1.00
Cube	0.81
Cylinder	
$h = d$	0.87
$h = 5d$	0.70
$h = 10d$	0.58
Disks	
$h = d/3$	0.76
$h = d/6$	0.60
$h = d/10$	0.47
Old beach sand	As high as 0.86
Young river sand	As low as 0.53
Average for various types of sand	0.75
Crushed solids	0.5–0.7
Granular particles	0.7–0.8
Wheat	0.85
Raschig rings	0.26–0.53
Berl saddles	0.30–0.37
Nickel saddles	0.14

[a] Data from Brown (1950), and from geometrical considerations.

Fig. 6.1. An irregular shaped particle.

Let us define the particle size d_p in such a way as to be useful for flow and pressure drop purposes. With this in mind, how we evaluate d_p depends on the type of instrument available to measure size.

1. For *large particles* (> 1 mm) we determine the size

- by weighing a known number of particles, if their density is known, or
- by fluid displacement of a known number of particles, if the particles are nonporous, or
- by calipers or micrometer, if the particles are regular in shape.

From these measurements first calculate the equivalent spherical diameter, defined as follows:

$$d_{\text{sph}} = \left(\begin{array}{c} \text{diameter of sphere} \\ \text{having the same volume} \\ \text{as the particle} \end{array} \right) \tag{6.2}$$

from which the particle size d_p is given as

$$d_p = \phi \cdot d_{\text{sph}} \tag{6.3}$$

where ϕ is measured directly or is estimated from Table 6.1.

2. For *intermediate sizes* screen analysis is the most convenient way of measuring the size. Each manufacturer has his own designated sizes, so be sure you know whose screens you are using. As an example, the Tyler standard screens, widely used, have sizes shown in Table 6.2. The screen size d_{scr} is then the mean of the aperture of the screen which just lets the particle through and the screen on which they are resting. For example, particles passing through a 150-mesh screen but resting on a 200-mesh screen would be called $-150 + 200$ mesh particles and would have a screen size

$$d_{\text{scr}} = \frac{104 + 74}{2} = 89 \, \mu\text{m}$$

Table 6.2. Tyler Standard Screen Sizes

Mesh number (number of wires/in)	Aperture, μm (opening between adjacent wires)	Mesh number (number of wires/in)	Aperture, μm (opening between adjacent wires)
3	6680	35	417
4	4699	48	295
6	3327	65	208
8	2362	100	147
10	1651	150	104
14	1168	200	74
20	833	325	53
28	589	400	38

Unfortunately there is no general relationship between d_{scr} and d_p. The best way we can say is

- for irregular particles with no seeming longer or shorter dimension take

$$d_p \cong \phi d_{scr} \qquad (6.4)$$

- for irregular particles with one somewhat longer dimension, but with length ratio not greater than 2 : 1, eggs for example,

$$d_p \cong d_{scr} \qquad (6.5)$$

- for irregular particles with one somewhat shorter dimension, but with length ratio not less than 1 : 2

$$d_p \cong \phi^2 d_{scr} \qquad (6.6)$$

- for extreme needle or fiberlike particles

$$d_p > d_{scr}$$

- for very flat pancake-shaped particles

$$d_p < \phi^2 d_{scr}$$

- for these flat-plate or needlelike particles it may be better to approximate the shape with idealized cylinders or disks.

3. For *very small particles* (< 40 μm) we rely on indirect methods such as sedimentation or Brownian motion to measure the particle size.

4. For *a size distribution of particles*, such as shown in Fig. 6.2, there are many ways of defining an average size. Since it is the surface of the

Fig. 6.2. Histogram representing the size distribution of particles in a packed bed.

particle which gives frictional resistance to flow we want a surface-average particle size. Thus we define this mean size as

$$\bar{d}_p = \left(\begin{array}{c} \text{the single size of particle which would have the same total} \\ \text{surface area as the size mixture in question—same total} \\ \text{bed volume and same bed voidage in both cases} \end{array} \right)$$

This definition leads to the following simple expression

$$\bar{d}_p = \frac{1}{\displaystyle\sum_{\substack{\text{all size} \\ \text{cuts}}} \left(\frac{x_i}{d_{pi}} \right)} \tag{6.7}$$

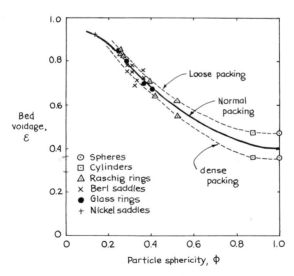

Fig. 6.3. The voidage increases as the sphericity decreases for randomly packed beds of uniform particles. [Adapted from Brown et al. (1950).]

Fig. 6.4. The wall region of a packed bed has an increased voidage. [From Leva (1957).]

C. Bed Voidage, ε

Typical voidage data for packed beds are shown in Figs. 6.3 and 6.4. Also, many equations and correlations have been proposed to give bed voidage. however, if it is important to get a good value for design it is best to measure the voidage of the packed bed directly. This is not that hard to do.

II. FRICTIONAL LOSS FOR PACKED BEDS

For fluid flowing through a packed bed of solids as shown in Fig. 6.5, the characteristic Reynolds number is defined as

$$\text{Re}_p = \frac{d_p u_0 \rho}{\mu} \tag{6.8}$$

where u_0 is called the superficial velocity. This is the velocity that the fluid would have if the vessel contained no solids.

Fig. 6.5. Flow through a packed bed.

The frictional loss for flow through packed beds can be expressed as

$$\Sigma F = \frac{150(1-\varepsilon)^2 \mu u_0 L}{g_c \varepsilon^3 d_p^2 \rho_g} + \frac{1.75(1-\varepsilon) u_0^2 L}{g_c \varepsilon^3 d_p} \quad \left[\frac{J}{kg}\right] \quad (6.9)$$

$$\underbrace{\hspace{3cm}}_{\substack{\text{Viscous} \\ \text{losses}}} \qquad \underbrace{\hspace{2cm}}_{\substack{\text{Turbulent} \\ \text{losses}}}$$

Equation (6.9) is an expression proposed by Ergun (1952) based on the clever combination of the Kozeny–Carman equation for flow in the viscous region, and the Burke–Plummer equation for the turbulent region. This two-term expression fits the data well, is widely used, and is called, not surprisingly, the Ergun equation. When $Re_p < 20$ the viscous loss term dominates and can be used alone with negligible error. On the other hand, when $Re_p > 1000$ only the turbulent loss term need be used.

Let us represent this correlation on a friction factor versus the Reynolds number plot, somewhat as in pipe flow. For this define the friction factor as

$$f_f = \left(\frac{\text{frictional energy loss} \Big/ \text{kg of fluid}}{\text{kinetic energy loss} \Big/ \text{kg of fluid}}\right) = \frac{\varepsilon^3}{1-\varepsilon} \cdot \frac{\Sigma F g_c d_p}{u_0^2 L} \quad (6.10)$$

Fig. 6.6. Friction factor vs. Reynolds number for flow through packed beds.

The Ergun equation then becomes

$$f_f = 150\frac{1-\varepsilon}{Re_p} + 1.75 \qquad (6.11)$$

which is displayed in Fig. 6.6.

Either use equation (6.11) with equation (6.10), or use equation (6.9) directly to determine the frictional losses in packed beds.

III. MECHANICAL ENERGY BALANCE FOR PACKED BEDS

Consider the flow system of Fig. 6.7. A mechanical energy balance between points 1 and 5 of the system gives

$$\frac{g}{g_c}\Delta z + \Delta\left(\frac{u^2}{2g_c}\right) + \int\frac{dp}{\rho_g} + W_s + \Sigma F = 0 \qquad \left[\frac{J}{kg}\right] \qquad (6.12)$$

Consider the individual terms in this expression.

1. Friction term, ΣF. Because the packed-bed section has a very large interfacial area its frictional loss is usually much greater than that of the rest of the piping system. Thus, we often can consider its frictional loss alone, or

$$\Sigma F_{total} \cong \Sigma F_{packed\ section}$$

Fig. 6.7. Flow system which includes a packed bed.

2. **The potential energy term, $g\Delta z / g_c$.** For gases this is usually negligible, but for liquids this can be an important term in the mechanical energy balance.

3. **The flow work term in systems with small density changes, $\int dp/\rho$.** When the density of the fluid does not vary much as it passes though the packed bed one can use an average density in the system. Thus

$$\int \frac{dp}{\rho} \cong \frac{\Delta p}{\bar{\rho}} \qquad (6.13)$$

This condition is satisfied for all liquids and for gases where the relative pressure variation is less than 10%, or where

$$\Delta p < 0.1\bar{p}$$

4. **The flow work term for gases experiencing large density changes, $\int dp/\rho$.** When the frictional pressure drop is large, meaning when $\Delta p > 0.1\bar{p}$, one should properly account for the change of density with pressure. Thus, combining equation (9) with the mechanical energy balance of equation (6.12), dropping the kinetic and potential energy terms, dividing by u_0^2, and introducing the superficial mass velocity $G_0 = u_0\rho_g$, gives

$$\int \frac{\rho_g \, dp}{G_0^2} + \frac{150(1 - \varepsilon)^2 \mu L}{g_c \varepsilon^3 d_p^2 G_0} + \frac{1.75(1 - \varepsilon)L}{g_c \varepsilon^3 d_p} = 0 \qquad (6.14)$$

If the gas expands reversibly as it flows through the packed bed it would cool. Here, however, flow is not reversible, friction causes heating, so it probably is best to assume isothermal flow. With this and the ideal gas assumption integration from point 3 to point 4 gives

$$\frac{g_c(mw)}{2G_0^2 RT}\left(p_4^2 - p_3^2 \right) + \frac{150(1 - \varepsilon)^2 \mu L}{\varepsilon^3 d_p^2 G_0} + \frac{1.75(1 - \varepsilon)L}{\varepsilon^3 d_p} = 0 \qquad (6.15)$$

Note how this expression parallels equation (3.21).

This isothermal packed-bed expression is the one to use in gas flow systems experiencing large pressure changes.

5. The kinetic energy term, $(\Delta u^2 / 2g_c)$. This is usually negligible for both liquids and gases because one rarely gets to very high velocities in packed beds. However, if one has to include this factor one should use the approach of Chapter 3 [see equations (3.12) and (3.21)] and add the correct term to equation (6.15).

6. The work term, W_s. Taking an energy balance about the whole system from point 1 to point 5 gives the shaft work directly. Alternatively, knowing p_1 and p_2, a mechanical energy balance about the ideal compressor alone (KE, PE, and ΣF all ignored) gives the energy actually received by the flowing fluid as

$$-W_s = \int_1^2 \frac{dp}{\rho_g} \qquad \left[\frac{\text{J}}{\text{kg}}\right] \tag{6.16}$$

For liquids and for gases experiencing a small fractional change in pressure (or $\Delta p < 0.1\ \bar{p}$) this expression simplifies to give

$$-W_s = \frac{\Delta p}{\bar{\rho}} \tag{6.17}$$

For gases experiencing a high fractional change in pressures (or $\Delta p > 0.1\ \bar{p}$), equation (6.16) is integrated to give equations (1.9) to (1.14).

7. Comments. Although most workers will agree that the form of the Ergun equation reasonably represents the frictional loss in packed beds, some have suggested that we do not trust too strongly the values of the Ergun constants, 150 and 1.75.

For example, in blast furnace work with very large particles, Standish and Williams (1975) find that the Ergun constants should both be doubled or tripled. Again, Macdonald et al. (1979) recently suggested that the 150 be replaced by 180, and that the 1.75 be replaced by 1.8 for smooth particles and 4.0 for very rough particles.

However, keeping in mind that most of the data points vary by a factor of two from the mean, we prefer, for the time being, to stay with the original Ergun equation with its considerable backing of experimental verification.

REFERENCES

G. G. Brown et al., Unit Operations, Chapter 16, Wiley, New York (1950).
S. Ergun, Fluid flow through packed columns, Chem. Eng. Prog. **48**, 89 (1952).
M. Leva, Variables in fixed bed systems, Chem. Eng., p. 263 (August 1957).

I. F. Macdonald, M. S. El-Sayed, K. Mow, and F. A. L. Dullien, Flow through porous media—the Ergun equation revisited, *Ind. Eng. Chem. Fundam.* **18**, 199 (1979).

N. Standish and I. D. Williams, *Proc. of a Blast Furnace Aerodynamics Symp.*, Aus. I.M.M., Wollongong, NSW, Australia (1975).

X. Schlünder, *Chemical Reaction Engineering Reviews—Houston*, ACS Symposium Series No. 72, p. 110 (1978).

EXAMPLE 6.1. A LABORATORY PACKED-BED EXPERIMENT

On the second floor of the laboratory of the East China Institute of Chemical Technology is a 0.22-m-i.d. glass tube packed to a depth of 1 m with 10-mm spheres as shown below. What will be the superficial velocity of water at 20°C through the bed if the water level is kept 3 m above the top of the bed?

Solution

We can choose to take the mechanical energy balance between various pairs of points in the system; thus, between points 1 and 3, 1 and 4, 2 and 3, or 2 and 4. However, in all cases the packed-bed section should be included, because that is where most of the frictional loss occurs. Let us take the balance between points 1 and 4, in which case $\Delta p/\rho$ and the kinetic energy terms drop out. Thus we obtain

$$\frac{g\Delta z}{g_c} + \cancel{\frac{\Delta u^2}{2g_c}}^{=0} + \cancel{\frac{\Delta p}{\rho}}^{=0} + \cancel{W_s}^{=0} + \Sigma F = 0 \qquad \text{(i)}$$

With no pump or turbine in the system $W_s = 0$, so we are left with

$$\Sigma F + \frac{g}{g_c}(z_4 - z_1) = 0$$

Now the frictional loss from points 1 to 4 consists of the resistance of the packed-bed section 2–3 and the empty pipe sections 1–2 and 3–4. However, it is reasonable to assume that the resistance of the empty pipe sections is negligible compared to that of the packed section. Thus, applying the Ergun equation (6.9), and estimating the bed voidage to be 0.38 from Figs. 6.3 and 6.4, we obtain

$$\frac{150(1 - 0.38)^2(10^{-3})u_0(1)}{(1)(0.38)^3(0.01)^2(1000)} + \frac{1.75(1 - 0.38)u_0^2(1)}{(1)(0.38)^3(0.01)} + \frac{(9.8)}{(1)}(0 - 3) = 0 \quad \text{(ii)}$$

or

$$10.51\, u_0 + 1977\, u_0^2 - 29.4 = 0 \tag{iii}$$

Solving the quadratic gives

$$u_0 = \boxed{0.119\ \text{m/s}}$$

NOTE: The above solution uses the complete Ergun equation. Had we guessed that the turbulent losses dominated, we would have dropped the linear term in the above quadratic, equation (iii), giving

$$u_0 = \left(\frac{29.4}{1977}\right)^{1/2} = \boxed{0.122\ \text{m/s}}\ (2.5\%\ \text{high})$$

And then a Reynolds number check would show

$$\text{Re}_p = \frac{d_p u_0 \rho}{\mu} = \frac{(0.01)(0.122)(1000)}{10^{-3}} = 1222$$

which justifies the assumption of turbulent flow.

The solution to this problem is quite sensitive to the value chosen for the bed voidage because the voidage, in effect, appears to the fourth and fifth powers in equation (ii). So if we had chosen $\varepsilon = 0.42$, instead of $\varepsilon = 0.38$, we would have found for equation (i)

$$6.81\, u_0 + 1370\, u_0^2 - 29.4 = 0$$

or

$$u_0 = \boxed{0.144\ \text{m/s}}\ (21\%\ \text{high})$$

This large uncertainty in reading ε from Figs. 6.3 and 6.4 is a reflection of the large variation of voidage obtainable in a packed bed. How you pack the bed, whether you shake and tap as you pour the solids in, and so on, can give widely different values for voidages.

In the setups above, with any value for x and y, the driving force remains at 3 m, the resistance is unchanged—still at 1 m of packed bed—so the solution is the same as presented above.

PROBLEMS ON PACKED BEDS

6.1. I predict that "Happy Bananas" will soon sweep the country. They will be produced by passing readily absorbed nitrous oxide (laughing gas) through a packed bed of green but full-grown Central American bananas. In developing this process we will need to know the pressure drop in these beds of bananas. To achieve this, estimate the effective banana size d_p from the geometric considerations shown here:

6.2. We plan to pack a tower with Raschig rings of dimensions shown below. Determine the effective particle size d_p of this packing material.

Note: The advantage of these specially designed packings is that they give small particle behavior coupled with large voidage—and high voidage gives a low pressure drop.

6.3. For pressure drop purposes a packed bed of spheres of what size would behave like a randomly packed mixture of equal weights of 1 mm and 2 mm spheres? Assume the same voidage in both beds.

6.4. Air at about 20°C and 1 atm, passes upward through a fixed bed ($L = 0.8$ m, $d_{bed} = 0.1$, $\varepsilon = 0.4$) of spherical particles ($\rho_s = 3000$ kg/m^3, $d_p = 10$ mm) at a superficial velocity of 1.5 m/s. Find the pressure drop across the bed.

6.5. Water flows downward through a tube inclined 30° from the horizontal and packed 10 m in length with metal spheres ($d_p = 1$ mm, $\rho_s = 5200$ kg/m^3, $\varepsilon = 0.34$). At a particular flow velocity the pressure is 3 atm at both ends of the bed.

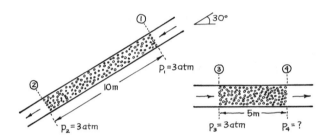

The tube is now tipped horizontally, the packed length is reduced to 5 m, and water flows at the same rate through the bed. If the pressure at the entrance of the bed is 3 atm, what is it at the exit of the bed?

6.6. Water flows downward through a vertical tube packed 10 m deep with metal spheres ($d_p = 1$ mm, $\rho_s = 5200$ kg/m^3, $\varepsilon = 0.34$). At a particular flow velocity the pressure just above the bed is 3 atm. Just below the bed it is also 3 atm.

The tube is now tipped 45° from from the vertical and water at the same flow rate flows upward through the bed. If the pressure just at the entrance is 3 atm, what is it at the exit of the bed?

6.7. For the regenerator of Example 18.1, air is compressed and then passed at a mass velocity of 4.8 kg/m^2 s through a vessel 54.5 m high, 1 m^2 in cross section, and filled with close to spherical packing ($d_p = 0.05$ m, $\varepsilon = 0.4$). Air leaves the packed bed at 20°C and 1 atm. What size of ideal compressor will provide this air flow rate?

6.8. *Gas flow in packed-bed catalytic reactors.* Packed beds of catalyst with wall cooling are widely used in the process industry to effect strongly exothermic reactions; for design purposes the engineer must be able to develop a model which reasonably approximates what is happening in these reactors. Such a model should consider a number of phenomena: the flow of generated heat outward to the cooling walls, the dispersion of flowing material radially and axially, and also the velocity distribution of the flowing gas.

This flow can be pictured in various ways, from the very simple to the more real but complex. For example, the simplest model assumes plug flow of gas, which means that all fluid elements move at exactly the same velocity through the packed bed with no overtaking. The second-stage model assumes plug flow of gas with small random velocity fluctuations superimposed. This is called the axial dispersion model.

Now it is well known that the voidage by the walls of a packed bed is higher than in the main body of the bed, so the third-stage model visualizes two flow regions, a central core of lower voidage surrounded by an annular wall region of higher voidage and one-particle diameter in thickness. Plug flow is assumed to occur in each region but with a higher velocity in the high-voidage region.

More precise models may try to incorporate the true velocity profile in the packed bed, a profile which in fact deviates rather significantly from plug flow; however, this profile is not reliably known today. Schlünder (1978) gives a good accounting of the state of knowledge of the factors involved in the proper modeling of fixed bed reactors.

To go back to the third-stage model, if the voidage in the main body of the packed bed of spherical catalyst particles is 0.36 and if it is 0.5 in the wall region, find the velocity ratio of fluid flowing in these regions for:

(a) very small particles and slow flow;

(b) for very large particles and high flow rate.

If the packed bed reactor is 8 particles in diameter, what fraction of the fluid flows up the wall region:

(c) for very small particles and slow flow?

(d) for very large particles and high flow rate?

6.9. In one design of a direct-contact crossflow gas–solid heat exchanger, the solids are conveyed horizontally on a screen conveyor while gas percolates upward through the solid. Suppose that hot solids ($\bar{d}_p = 10$ mm, $\varepsilon = 0.4$) are conveyed at 0.2 m/s in a layer 0.2 m thick while cold air flows upward through the solid mass from a high-pressure chamber (pressure is 2 kPa above atmospheric) to atmospheric, the whole exchanger being at an average temperature of 100°C. Find the direction of air flow through the layer of solid, and give this as the angle from the vertical θ, as shown in Fig. 6.P9.

6.10. The critcial step in Motorola's proposed process for producing ultra pure solar-cell grade silicon is the reaction

$$\text{Si}\left(\begin{array}{c}\text{commercial grade,}\\ \text{solid, 98\% pure}\end{array}\right) + \text{SiF}_4\left(\begin{array}{c}\text{gas,}\\ \text{pure}\end{array}\right) \rightarrow 2\,\text{SiF}_2\left(\begin{array}{c}\text{gas,}\\ \text{pure}\end{array}\right) \cdots \text{exothermic}$$

The SiF$_2$ is then decomposed elsewhere to produce pure silicon. Thermody-

namics says that at the chosen operating temperature of 1350 K the reaction
should only become appreciable when the pressure falls below 100 Pa. The
reactor will be a vertical 0.4-m-i.d. tube packed with 6-mm silicon particles
($\varepsilon = 0.5$). Reactant gas (pure SiF_4) will enter the reactor at 2000 Pa and 1350
K, and 1.2 m downstream we expect the pressure to have dropped to 100 Pa.
From here on reaction takes off, the temperature rises, and conversion to
product is soon complete. Estimate the production rate of silicon (kg/hr)
obtainable from this reactor.

  Data:

$$\mu_{SiF_4} = 4 \times 10^{-5} \text{ kg/m s (estimated)}$$

$$(mw)_{Si} = 0.028$$

$$(mw)_{SiF_4} = 0.104$$

$$(C_p/C_v)_{SiF_4} = 1.15 \text{ (estimated)}$$

Chapter 7   FLOW IN
FLUIDIZED BEDS

I. THE FLUIDIZED STATE

Suppose we progressively increase the velocity of fluid which is flowing upward through a batch of solids. The frictional resistance increases, and eventually a point is reached where the resistance just equals the weight of solids. At this point the solids become suspended—in other words, they

Fig. 7.1. Transition from a packed to a fluidized bed as the velocity of upflowing fluid is raised.

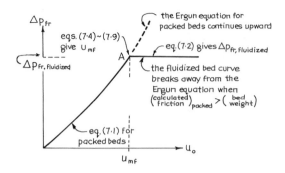

Fig. 7.2. Frictional loss in the packed and in the fluidized state.

become "fluidized"—and the superficial velocity of fluid needed to just fluidize the solids is called the minimum fluidizing velocity u_{mf}. As the flow rate of fluid is increased beyond this point a liquid fluidized bed keeps expanding, while a gas bed expands only slightly. This progression is shown in Fig. 7.1. The fluidized state has many desirable liquidlike properties. For example, we can easily move solids about just by pumping or by gravity flow.

The progression from packed bed to fluidized bed is best followed on a pressure drop vs. velocity graph, as shown in Figs. 7.2 and 7.3. The packed bed behavior, shown on the left of these figures is reasonably represented by the Ergun equation (6.9)

$$\Delta p_{fr} = \rho_g \Sigma F = \frac{150}{g_c} \frac{(1-\varepsilon)^2}{\varepsilon^3} \frac{\mu u_0 L}{d_p^2} + \frac{1.75}{g_c} \frac{1-\varepsilon}{\varepsilon^3} \frac{\rho_g u_0^2 L}{d_p} \qquad \left[\frac{N}{m^2} = Pa\right]$$

$$(7.1)$$

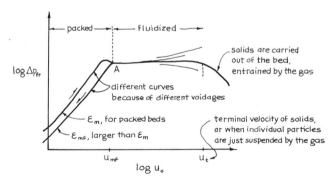

Fig. 7.3. The pressure drop vs. velocity curve in more detail and on log–log coordinates.

where the characteristic particle size d_p is given by equations (6.3)–(6.7). From this equation we see that

$$\Delta p_{fr} \propto u_0 \quad \text{(at low velocity)}$$

$$\Delta p_{fr} \propto u_0^2 \quad \text{(at high velocity)}$$

Point A in Figs. 7.2 and 7.3 represents the just fluidized bed; thus, it is at the minimum fluidizing velocity. To the right of point A the bed is well fluidized. In the following sections we will find

- the pressure drop and pumping power needed to fluidize a bed of solids.
- the minimum fluidizing velocity, given by point A of Figs. 7.2 and 7.3.

II. FRICTIONAL LOSS AND PUMPING REQUIREMENT NEEDED TO FLUIDIZE A BED OF SOLIDS

Consider the setup of Fig. 7.4. Solids will just fluidize ($u_0 = u_{mf}$, $\varepsilon = \varepsilon_{mf}$) when

$$\begin{pmatrix} \text{total upward} \\ \text{drag force by} \\ \text{fluid on particles} \end{pmatrix} = \begin{pmatrix} \text{total weight} \\ \text{of solids} \\ \text{in the bed} \end{pmatrix}$$

or

$$\begin{pmatrix} \text{frictional} \\ \text{pressure drop} \end{pmatrix} \begin{pmatrix} \text{cross-sectional} \\ \text{area of bed} \end{pmatrix} = \begin{pmatrix} \text{volume of} \\ \text{bed} \end{pmatrix} \begin{pmatrix} \text{fraction} \\ \text{consisting} \\ \text{of solids} \end{pmatrix} \begin{pmatrix} \text{specific} \\ \text{weight} \\ \text{of solids} \end{pmatrix}$$

Fig. 7.4

or, in symbols,

$$\Delta p_{fr} \cdot A_t = A_t L_{mf}\left(1 - \varepsilon_{mf}\right)\left[\left(|\rho_s - \rho_g|\right)\right]\left(g/g_c\right)$$

Writing the mechanical energy balance of equation (6.12) between points 1 and 2 of Fig. 7.4 and ignoring kinetic energy effects, gives for unit cross-sectional area of bed

$$\Delta p_{fr} = \rho_g \Sigma F = -\left(\Delta p + \frac{\rho_g g \Delta z}{g_c}\right) = L_{mf}\left(1 - \varepsilon_{mf}\right)\left(|\rho_s - \rho_g|\right)\frac{g}{g_c} \quad (7.2)$$

Frictional
pressure drop

Ignore
for gases

$= W/A_t$

The pumping requirement is found by writing the mechanical energy balance between points 0 and 2. This gives

$$-W_s = \Sigma F_{\text{for the bed}} = \frac{\Delta p_{fr}}{\rho_g} \qquad \left[\frac{J}{kg}\right] \qquad (7.3)$$

Equations (1.9)–(1.14) deal further with pumping requirements.

In more realistic situations the frictional loss of the air distribution plate holding up the solids, the cyclone separator or bag filters above the bed, and the other equipment in the line between points 1 and 2, must be accounted for. Thus, in place of equation (7.3), we should more properly write

$$-W_s = \Sigma F_{\text{for the bed}} + \Sigma F_{\substack{\text{distributor} \\ \text{plate}}} + \Sigma F_{\substack{\text{cyclone} \\ \text{separator}}} + \text{etc.} \qquad (7.3a)$$

At flow velocities higher than u_{mf} the frictional loss per unit of fluid passing through the bed remains practically unchanged, as shown in Figs. 7.2 or 7.3; however, the pumping power needed will change in proportion to the fluidizing velocity u_0.

III. MINIMUM FLUIDIZING VELOCITY, u_{mf}

The intersection of equations (7.1) and (7.2) represents the condition where the solids just fluidize (point A of Fig. 7.2). So combining these two equations gives the following expression for finding the minimum fluidizing velocity

$$\left. \begin{array}{l} \dfrac{1.75}{\varepsilon_{mf}^3}\left(\dfrac{d_p u_{mf}\rho_g}{\mu}\right)^2 + \dfrac{150(1-\varepsilon_{mf})}{\varepsilon_{mf}^3}\left(\dfrac{d_p u_{mf}\rho_g}{\mu}\right) = \dfrac{d_p^3 \rho_g(|\rho_s - \rho_g|)g}{\mu^2} \\[3em] \text{or} \\[1em] \underbrace{\dfrac{1.75}{\varepsilon_{mf}^3}\text{Re}_{p,mf}^2}_{\substack{\text{Ignore this}\\\text{term for}\\\text{small }d_p}} + \underbrace{\dfrac{150(1-\varepsilon_{mf})}{\varepsilon_{mf}^3}\text{Re}_{p,mf}}_{\substack{\text{Ignore this}\\\text{term for}\\\text{large }d_p}} = \underset{\substack{\text{Archimedes}\\\text{number}}}{\text{Ar}} \end{array} \right\} \quad (7.4)$$

In the special case of very small or very large particles the above expression simplifies as follows: for *very small particles*

$$u_{mf} = \frac{d_p^2(|\rho_s - \rho_g|)}{150\,\mu} \cdot \frac{g\varepsilon_{mf}^3}{1-\varepsilon_{mf}} \qquad \text{Re}_{p,mf} < 20 \qquad (7.5)$$

and for *very large particles*

$$u_{mf}^2 = \frac{d_p(|\rho_s - \rho_g|)}{1.75\,\rho_g} \cdot g\varepsilon_{mf}^3 \qquad \text{Re}_{p,mf} > 1000 \qquad (7.6)$$

If ε_{mf} and/or sphericity ϕ (which enters in d_p) are not known, we can use the modification of these equations suggested by Wen and Yu (1966). For a variety of systems they found that

$$\frac{1}{\phi\varepsilon_{mf}^3} \cong 14 \quad \text{and} \quad \frac{1-\varepsilon_{mf}}{\phi^2\varepsilon_{mf}^3} \cong 11$$

Replacing these expressions in equation (7.4) and combining with equation (6.3) then gives

$$\frac{d_{\text{sph}} u_{mf}\rho_g}{\mu} = \left[(33.7)^2 + \frac{0.0408\,d_{\text{sph}}^3\rho_g(|\rho_s - \rho_g|)g}{\mu^2}\right]^{1/2} - 33.7 \quad (7.7)$$

For small particles this expression simplifies to give

$$u_{mf} = \frac{d_{\text{sph}}^2(|\rho_s - \rho_g|)g}{1650\,\mu} \qquad \text{Re}_{mf} < 20 \qquad (7.8)$$

and for large particles

$$u_{mf}^2 = \frac{d_{\text{sph}}(|\rho_s - \rho_g|)g}{24.5\,\rho_g} \qquad \text{Re}_{mf} > 1000 \qquad (7.9)$$

where d_{sph} is related to the screen diameter d_{scr} by equations (6.3) and either (6.4), (6.5), (6.6), or (6.7).

For more on the flow characteristics of fluidized beds see Kunii and Levenspiel (1979).

REFERENCES

C. Y. Wen and Y. H. Yu, A generalized method for predicting the minimum fluidization velocity, *AIChE J.* **12**, 610 (1966).

D. Kunii and O. Levenspiel, *Fluidization Engineering*, Chapter 3, Krieger, Melbourne, FL (1979).

EXAMPLE 7.1. POWER TO RUN A FLUIDIZED INCINERATOR FOR MUNICIPAL GARBAGE

Room temperature air (20°C) is to be compressed and fed to a fluidized incinerator which operates at high temperature and uses sand as a carrier solid. Find the power requirement of the compressor if the unit is to operate at ten times the minimum fluidizing velocity.

Data:

• Incinerator: 3-m-i.d., slumped bed height = 0.56 m

$$\varepsilon_m = 0.36, \ \varepsilon_{mf} = 0.44, \ \varepsilon_f = 0.54$$

• Solids: $-28 + 35$ mesh sand, $\rho_s = 2500 \text{ kg/m}^3$, $\phi = 0.875$.
• Temperature of the bed = 850°C. Because of its remarkably good heat distribution characteristics, a well-fluidized bed is close to uniform in temperature, so assume that the air reaches 850°C as soon as it enters the bed.
• Take $\mu_{\text{air},850°C} = 4.5 \times 10^{-5} \text{ kg/m s}$.
• Efficiency of the compressor and motor = 70%.
• Ignore the pressure drop of the distributor and of the cyclones.

Solution

Our strategy will be to determine, in turn, the pressure drop through the bed, the gas velocity through the bed, and finally, the power requirement. But let us first evaluate a few needed quantities.

$$L_{mf} = \frac{L_m(1 - \varepsilon_m)}{1 - \varepsilon_{mf}} = \frac{0.56(1 - 0.36)}{1 - 0.44} = 0.64 \text{ m}$$

$$d_{scr} = \frac{417 + 589}{2} = 503 \; \mu\text{m, from Table 6.2}$$

$$d_p = \phi d_{scr} = 0.875(503) = 440 \; \mu\text{m, from equation (6.4)}$$

Calculate Δp_{fr} from equation (7.2)

$$\Delta p_{fr} = \rho_g \Sigma F = L_m(1 - \varepsilon_m)(\rho_s - \rho_g)\frac{g}{g_c}$$

$$= 0.56(1 - 0.36)(2500)9.8 = 8781 \text{ Pa}$$

$$p_2 = 101,325 + 8781 = 110,106 \text{ Pa}$$

and

$$\bar{\rho}_{g,\text{in bed}} = \frac{(mw)\bar{p}}{RT_3} = \frac{0.0289\left(\dfrac{110,106 + 101,325}{2}\right)}{(8.314)(1123)}$$

$$= 0.327 \text{ kg/m}^3$$

Calculate u_{mf} from equation (7.1)

$$\Delta p_{fr} = \frac{150\left(1 - \varepsilon_{mf}\right)^2 \mu u_{mf} L_{mf}}{g_c \varepsilon_{mf}^3 d_p^2} + \frac{1.75\left(1 - \varepsilon_{mf}\right)\bar{\rho}_g u_{mf}^2 L_{mf}}{g_c \varepsilon_{mf}^3 d_p}$$

Replacing known values gives

$$8781 = \frac{150(1 - 0.44)^2(4.5 \times 10^{-5})u_{mf}(0.64)}{1(0.44)^3(0.00044)^2}$$

$$+ \frac{1.75(1 - 0.44)(0.327)u_{mf}^2(0.64)}{1(0.44)^3(0.00044)}$$

or

$$8781 = 82{,}148\, u_{mf} + 5472\, u_{mf}^2$$

and solving gives

$$u_{mf} = 0.106 \text{ m/s}$$

Hence the velocity of gas entering the bed at 850°C and 110 kPa is

$$u_0 = 10\, u_{mf} = 1.06 \text{ m/s}$$

NOTE: Instead of finding Δp_{fr} and then u_{mf}, we could have found u_{mf} directly from equation (7.4). It will give the same answer.

Calculate the power requirement from equations (1.12) and (1.14) but first determine the molar flow rate of gas from the conditions of the gas just entering the bed. Thus

$$\dot{n} = \frac{p_3 \dot{v}_3}{RT_3} = \frac{(110106)(1.06)(\pi/4)(3)^2}{(8.314)(1123)} = 88.5 \text{ mol/s}$$

and now an energy balance about the compressor gives

$$-\dot{W}_s = \frac{k}{k - 1} \cdot \frac{\dot{n}RT_1}{\eta}\left[\left(\frac{p_2}{p_1}\right)^{(k-1)/k} - 1\right]$$

$$= \frac{1.4}{1.4 - 1}\frac{(88.5)(8.314)(293)}{0.7}\left[\left(\frac{110{,}106}{101{,}325}\right)^{(1.4-1)/1.4} - 1\right]$$

$$= 25902 \text{ J/s} = 26 \text{ kW}$$

PROBLEMS ON FLUIDIZED BEDS

7.1. Air at 20°C passes upward through a bed (2 m high, 0.5 m i.d., $\varepsilon_m = 0.4$, $\varepsilon_{mf} = 0.44$) of limestone ($d_p = 2$ mm, $\rho_s = 2900$ kg/m^3). What inlet air pressure is needed to fluidize the solids (outlet pressure = 1 atm)?

7.2. Air at 1.1 atm and 20°C filters upward through a 10-m high bed of 1-mm solid silver spheres ($\varepsilon_{mf} = 0.4$) and discharges to the atmosphere. Ten meters!! What a waste! Silver is valuable, so let's sell some of it. No one will know as long as the remaining solids remain as a fixed bed. What height of silver should we leave behind so that the bed won't fluidize?

7.3. We plan to pass air upward through a bed of solids resting on a screen. Will the solids fluidize?
Data:

Solids: $\rho_s = 3000$ kg/m^3, $d_p = 2$ mm, $\varepsilon_m = 0.36$

Bed: 2 m high, 0.5-m-i.d.

Air: $p_{inlet} = 130$ kPa, $p_{outlet} = 100$ kPa, $T = 100$°C

7.4. Air at 150 kPa and 20°C will flow upward at a superficial velocity of 1 m/s through a bed of solids.
(a) Will the solids fluidize?
(b) Find the outlet air pressure for this flow rate.
Data:

Solids: $\rho_s = 4500$ kg/m^3, $d_p = 1$ mm

Bed: $\varepsilon_m = 0.36$, 1 m high, 0.3-m-i.d.

7.5. Air enters ($u_0 = 1$ m/s, $p = 0.2$ MPa, $T = 293$ K) and passes upward through a bed of packed solid ($d_p = 1$ mm, $\varepsilon_m = 0.4$, $\rho_s = 9500$ kg/m^3, $L_m = 1$ m) sandwiched and kept in place between two screens having negligible resistance to air flow.
(a) What is the outlet pressure?
(b) The upper screen is removed. Now what is the outlet pressure?

7.6. Air at about 20°C and 1 atm passes upward through a fixed bed of solids.
    (a) At what superficial air velocity will the particles just fluidize?
    (b) At minimum fluidizing velocity what will be the pressure drop across
    the bed?
    Data:

$$\text{Solids: } d_{sph} = 15 \text{ mm}, \ \phi = 0.67, \ \rho_s = 3000 \text{ kg/m}^3$$

$$\text{Bed: } \varepsilon_m = \varepsilon_{mf} = 0.4, \text{ height} = 0.5 \text{ m, diameter} = 0.1 \text{ m}$$

7.7. Calculate u_{mf} for a bed of irregular particles (take $d_p = d_{scr} \cdot \phi_s$) fluidized by
    air at about 20°C and 1 atm.
    Data:

$$\text{Solids: } d_{scr} = 100 \ \mu\text{m}, \ \phi_s = 0.63, \ \rho_s = 5000 \text{ kg/m}^3$$

$$\text{Bed voidage: } \varepsilon_m = \varepsilon_{mf} = 0.6 \text{ (estimated)}$$

7.8. A U-tube is packed with 1-mm glass beads ($\rho_s = 2200 \text{ kg/m}^3$, $\varepsilon_m = 0.40$) and
    water flows through the system as shown below on the left. As the upper water
    level is changed so does the flow rate in the packed bed.
    (a) At what height H will the spheres just lift up and wash away?
    (b) What is the velocity of water when this just occurs?
7.9. A packed bed flow experiment is sketched below on the right. As the water
    flow rate is increased, the water level at point 1 rises and H increases.
    Eventually a point is reached when the plastic balls just fluidize downward.
    Determine the height H and the liquid flow rate when this should occur.

7.10. Here is a proposed ride for the Tokyo Disneyland. Crowd a group of 75 children into a Plexiglas cylinder 3-m-i.d. and fluidize them. Hands and feet flying, what fun it would be. And we'd call it "Typhoon", the Chinese word for "big wind."

(a) What air velocity would be needed to just fluidize the children?

(b) What size of compressor at 50% efficiency (to account for inefficiencies, distributor resistance, screen above the bed to catch the occasional blown-out child) is needed if $u_0 = 1.1\ u_{mf}$?

Data: Children weigh from 25 to 50 kg with 40 kg as a mean. For suitably dressed screaming children (crash helmets, padding and all) $\phi = 0.22$, $\rho = 800$ kg/m^3, and I suppose that $\varepsilon_m \cong 0.4$, $\varepsilon_{mf} = \varepsilon_f = 0.5$ are fair estimates of the void fraction between children.

7.11. Example 7.1 made the simplifying assumption that the pressure drop of the distributor plate and of the cyclone are relatively small and can be ignored. Relax this assumption and repeat Example 7.1 with the more realistic assumption that $\Delta p_{\text{distributor}} = 20\%\ \Delta p_{\text{bed}}$ and $\Delta p_{\text{cyclones}} = 10\%\ \Delta p_{\text{bed}}$.

7.12. *Production of ultrapure silicon.* Battelle Memorial Institute is developing a fluidized bed process for producing ultrapure solar cell grade silicon. The reaction is to proceed at 1200 K as follows

$$2Zn(g) + SiCl_4(g) \rightarrow 2ZnCl_2(g) + Si\downarrow$$

and the process looks somewhat as shown on the following page. If the discharge tube plugs up the process is ruined. To avoid this plugging, sticking, or bridging of solids we pass a bit of SiCl$_4$ gas *up* the tube to fluidize the solids. This will insure smooth downflow of solids.

(a) How would you be able to tell whether the solids in the discharge tube are fluidized or not?

(b) What gas flow up the tube would you need to keep the downflowing solids fluidized.

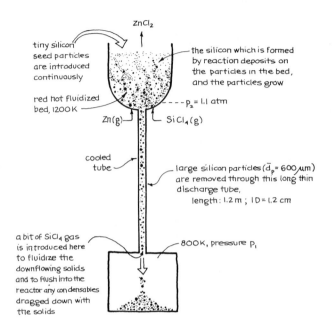

Data: Silicon particles in discharge tube:

$$d_p = 600 \ \mu\text{m}, \ \rho_s = 2200 \ \text{kg/m}^3, \ \varepsilon_{mf} = 0.4.$$

$SiCl_4$ gas at average conditions (1000 K) in the discharge tube:

$$(mw) = 0.170 \ \text{kg/mol}, \ \mu = 4 \times 10^{-5} \ \text{kg/m s}$$

7.13. *Cold model of an FBC* (*fluidized bed combustor*). Our laboratory has a cold model of a tube-filled fluidized furnace for the generation of steam from the combustion of coal. The unit consists of a vessel 1 m^2 in cross section, which fluidizes crushed rock, roughly 4 mesh in size, with a settled bed height of 0.5 m. Room temperature air (20°C) enters the compressor, but it is estimated that the bed itself will be hotter, roughly 40°C.

(a) Find the superficial gas velocity which will just fluidize the solids.

(b) We plan to operate the bed at four times the minimum fluidizing velocity. What size of pump–motor setup at 60% overall efficiency will be needed? Assume that the pressure drop through the distributor plate and the cyclone will be 20% and 10% of the bed, respectively.

Data:

$$\rho_s = 2900 \text{ kg/m}^3, \ d_{scr} = 5 \text{ mm}, \ \phi = 0.6$$

$$\varepsilon_m = 0.4, \ \varepsilon_{mf} = 0.5, \ \varepsilon_f = 0.6$$

$$\mu_{air}(40°C) = 1.96 \times 10^{-5} \text{ kg/m s}$$

7.14. Repeat the previous problem with one change: instead of using these large particles use roughly 28 mesh solids ($d_{scr} = 0.5$ mm).

Chapter 8 SOLID PARTICLES FALLING THROUGH FLUIDS

I. THE SPHERE

The forces acting on a sphere falling through a fluid (call it gas for convenience) are as follows:

$$\left(\begin{array}{c} \text{Net downward} \\ \text{force} \end{array} \right) = \left(\begin{array}{c} \text{weight of} \\ \text{particle} \end{array} \right) - \left(\begin{array}{c} \text{drag} \\ \text{force} \end{array} \right)$$

d_{sph}

frontal area
$= \frac{\pi}{4} d^2_{\text{sph}}$

In symbols

$$F = \frac{m}{g_c} \frac{du}{dt} = \left(\frac{\pi}{6} d^3_{\text{sph}} \right) \left(|\rho_s - \rho_g| \right) \frac{g}{g_c} - F_d \tag{8.1}$$

where the drag force

$$F_d = C_D \cdot \frac{\pi d_{sph}^2}{4} \frac{\rho_g u^2}{2g_c} \tag{8.2}$$

Drag coefficient

At the *terminal velocity du/dt* = 0, in which case equations (8.1) and (8.2) give

$$u_t = \left(\frac{4gd_{sph}(|\rho_s - \rho_g|)}{3\rho_g C_D} \right)^{1/2} \quad \text{or} \quad C_D = \frac{4gd_{sph}(|\rho_s - \rho_g|)}{3\rho_g u_t^2} \tag{8.3}$$

The drag coefficient in equation (8.3) has been found by experiment to be related to the particle Reynolds number at terminal velocity, defined as

$$\text{Re}_{sph,t} = \frac{d_{sph} u_t \rho_g}{\mu} \tag{8.4}$$

For the special case of *viscous flow of a sphere*, occurring when $\text{Re}_t < 1$, Stokes developed the following theoretical expression for the drag force

$$F_d = \frac{3\pi d_{sph} \mu u}{g_c} \tag{8.5}$$

Replacing in equation (8.4) then gives

$$\left. \begin{array}{l} u_t = \dfrac{(|\rho_s - \rho_g|) g d_{sph}^2}{18\mu} \\[2em] C_D = \dfrac{24}{\text{Re}_{sph,t}} = 24\left(\dfrac{\mu}{d_{sph} u_t \rho_g} \right) \end{array} \right\} \quad \text{good only when } \text{Re}_{sph,t} < 1 \quad (8.6)$$

For the *turbulent flow of a sphere*, $\text{Re}_t > 1$, no satisfactory theoretical expression for the drag force has been developed; consequently the frictional loss has to be found by experiment. These findings, shown as a C_D vs. Re_t relationship for $\phi = 1$ are given in Figs. 8.1 and 8.2. Figure 8.1 is useful

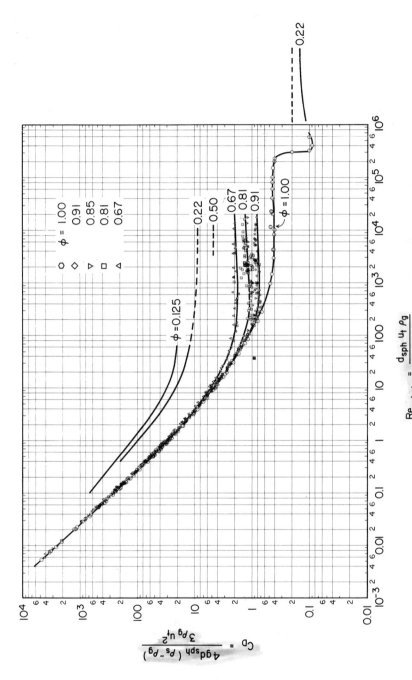

Fig. 8.1. Friction factor or drag coefficient *vs.* terminal Reynolds number for falling particles.

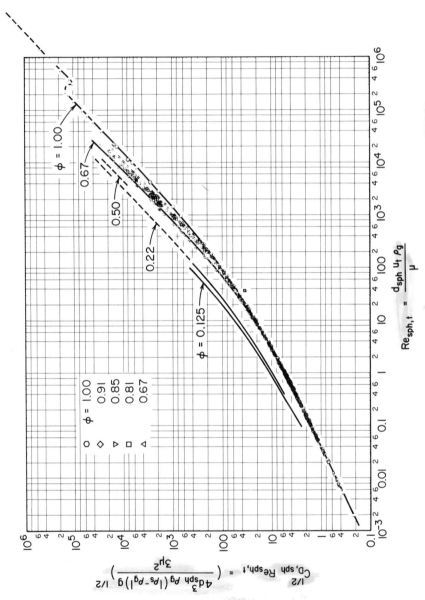

Fig. 8.2. Chart for determining the terminal velocity of particles falling through fluids.

for finding the frictional loss given u_t, while Fig. 8.2 is useful for finding u_t for given particles.

II. ANY SHAPE OF SOLIDS

For irregularly shaped falling solids the best we can do is find the frictional loss by experiment. These findings are shown in Figs. 8.1 and 8.2. Figure 8.2 is more useful for finding the terminal velocity of a given particle.

These figures were prepared by Cecen (1982) from data reported by Achenbach (1972), Pettijohn and Christiansen (1948), Schiller (1932), Schlichting (1979), Schmiedel (1928), and Wadell (1934).

REFERENCES

E. Achenbach, Experiments on the flow past spheres at very high Reynolds numbers, *J. Fluid Mech.* **54**, 565 (1972).

F. Cecen, M. S. Project, Chemical Engineering Department, Oregon State University, Corvallis (1982).

E. S. Pettyjohn and E. B. Christiansen, Effect of particle shape on free-settling rates of isometric particles, *Chem. Eng. Prog.* **44**, 157 (1948).

L. Schiller, *Hydro- und Aerodynamik Handbuch der Experimentalphysik*, Bd. IV, Teil 2, p. 335 (1932).

H. Schlichting, *Boundary Layer Theory*, Sixth Ed., p. 17, McGraw-Hill, New York (1979).

J. Schmiedel, Experimentelle Untersuchungen über die Fallbewegung von Kugeln und Scheiben in reibenden Flussigkeiten, *Phys. Z.* **29**, 593 (1928).

H. Wadell, The coefficient of resistance as a function of Reynolds number for solids of various shapes, *J. Franklin Inst.* **217**, 459 (1934).

EXAMPLE 8.1. SUING THE UNITED STATES FOR ITS MISBEHAVING VOLCANOS

On May 18, 1980, Mount St. Helens on the West Coast of the United States erupted catastrophically, spewing an ash plume to an altitude of 20 km. The winds then carried these millions of tons of particles, consisting mainly of silica ($\sim$ 70%), across the U.S., depositing a 2-cm layer on my campsite high (2 km) in the Rockies and 1000 km away. It started raining ash just 50 hr after the eruption, and although I left for home soon after, I had to breathe this contaminated air.

I am worried because I read on page 19 of the June 9, 1980 issue of *Chemical Engineering News* that silica particles smaller than 10 μm are respirable and can cause silicosis. No one told me this, I've sinced developed a cough, and being a normal American, I'm ready to sue the government for gross negligence in not warning me of this danger. But, of course, I will only do this if the particles are in the dangerous size range.

Please estimate the size of particles which settled on me at the start of this ash rain.

Data: Assume that ash particles consist of pure silica for which

$$\rho_s = 2650 \text{ kg/m}^3 \quad \text{and} \quad \phi = 0.6.$$

The average atmospheric conditions from 2 to 20 km are $T = -30°C$, $p = 40$ kPa, at which $\mu_{air} = 1.5 \times 10^{-5}$ kg/m s.

Solution

The problem statement does not specify what is meant by the words "particle size"—silica is quite nonspherical—so let us determine both the screen size and the equivalent spherical size. First we find

$$\rho_g = \frac{(mw)p}{RT} = \frac{(0.0289)(40,000)}{(8.314)(243)} = 0.5722 \text{ kg/m}^3$$

$$u_t = \frac{\text{distance fallen}}{\text{time}} = \frac{20,000 - 2000}{50 \times 3600} = 0.1 \text{ m/s}$$

Then

$$\text{Re}_{\text{sph},t} = \frac{d_{\text{sph}} u_t \rho_g}{\mu} = \frac{d_{\text{sph}}(0.1)(0.5722)}{1.5 \times 10^{-5}} = 3815 \, d_{\text{sph}} \tag{i}$$

and from equation (8.3)

$$C_{D,\text{sph}} = \frac{4 g d_{\text{sph}}(\rho_s - \rho_g)}{3\rho_g u_t^2} = \frac{4(9.8) d_{\text{sph}}(2650 - 0.57)}{3(0.5722)(0.1)^2} \tag{ii}$$

$$= 6.05 \times 10^6 \, d_{\text{sph}}$$

Now solve by trial and error using Fig. 8.1

Guess d_{sph}	$\text{Re}_{\text{sph},t}$ from equation (i)	$C_{D,\text{sph}}$ from equation (ii)	$C_{D,\text{sph}}$ from Fig. 8.1
1×20^{-5} m	0.038	60	670
10×10^{-5}	0.38	600	73
3.4×10^{-5}	0.13	206	205 (close enough)

Thus, d_{sph} = 34 μm, and for irregular particles with no particularly short or long dimension, equations (6.3) and (6.4) give

$$d_{scr} = \frac{d_p}{\phi} = d_{sph} = 34 \ \mu m$$

Conclusion. Either way you define particle size, you'd better not sue.

PROBLEMS ON FALLING OBJECTS

8.1. *Skydiving.* Indoor skydiving has come to Saint Simon, near Montreal, in the form of the "Aerodium," a squat vertical cylinder 12 m high and 6 m i.d., with safety nets at top and bottom. A DC-3 propeller which is driven by a 300-kW diesel engine blasts air upward through the Aerodium at close to 150 km/hr, while the "jumper" dressed in an air inflated jumpsuit floats, tumbles, and enjoys artificial free-fall in this rush of air without the danger of the real thing.

If an 80-kg, suited adult (density = 500 kg/m^3) in the spread-eagled position can hang suspended when the air velocity is 130 km/hr, what is his sphericity in this position? (Information from *Parachutist*, pg. 17, August 1981.)

8.2. The free-fall velocity of a very tiny spherical copper particle in 20°C water is measured by a microscope and found to be 1 mm/s. What is the size of the particle?

8.3. Who else could it happen to but "Bad Luck" Joe? He goes hunting, gets lost, fires three quick shots into the air, and gets hit squarely on the head by all three bullets as they come down. How fast were the bullets going when they hit him?

Data: Each bullet has a mass of 180 grains or 0.0117 kg, ϕ = 0.806 ρ_{bullet} = 9500 kg/m^3.

8.4. Water flows downward through a 1-m deep packed bed (ε = 0.4) of 1-mm plastic spheres (ρ_s = 500 kg/m^3). What head of water is needed to keep the spheres from floating upward?

8.5. Rutherford Arlington, a freshman who weighs 80 kg stripped, yens to streak with a difference, and stepping out of a balloon at 1000 m directly above Central Square during a noon revival meeting appeals to him. Imagine the impact—a real heavenly body entering their midst. At what speed would he join the faithful
    (a) if he curls up into a perfect sphere?
    (b) if he is spread-eagled? In this orientation, $\phi = 0.22$.

8.6. Find the upward velocity of air at 20°C which will just float a ping pong ball.
    Data: Nittaku 3 Star ping pong balls, used for the 37th World Championships in Tokyo in 1983, have a diameter of 37.5 mm and a mass of 2.50 g.

8.7. Lower Slobovia recently entered the space race with its own innovative designs. For example, the touchdown parachute of their lunar space probe, the "lunik," was ingeniously stored in the mouth of the braking rocket to save space. Unfortunately, instead of releasing the parachute, then firing the rocket, our intrepid spaceman, the "lunatic," first fired the rocket, which then used all its fuel to burn up the parachute—all this 150 km above the earth. When the Slob finally returned to earth he had a somewhat rough landing. At what speed do you estimate that he hit the ground?
    Information on the rocket: Volume = 5 m³, mass = 2.5 tons, and surface area = 20 m².

8.8. Rutherford Arlington, famed streaker, plans to use a helium balloon for the ascent prior to his spectacular free-fall (see Problem 8.5). To reach a height of 1000 m in 10 min, what size of balloon would he need?
    Data: The combined mass of Ruthy and his balloon = 120 kg, $\phi \cong 1$, $T = 20°C$, $\pi = 100$ kPa.
    (Problem prepared by Dan Griffith.)

8.9. Referring to the hypothetical ride called "Typhoon" for the Tokyo Disneyland (Problem 7.10) in which children are fluidized in a large Plexiglas cylinder, the only worry is that some small child may rise above his screaming fellows to get away from it all. To see if this is likely to occur, calculate the terminal velocity of a little Japanese child. See Problem 7.10 for additional data.

8.10. A bed of uniform particles fluidize in gas at u_{mf} and elutriate at u_t. Find the ratio of u_t/u_{mf}
    (a) for very fine solids ($u_{mf} = 1$ mm/s);
    (b) for very large solids ($u_{mf} = 1$ m/s).

This ratio, u_t/u_{mf}, is a good measure of the flexibility of a fluidized bed in accommodating variations in gas flow.

(c) Which size of solids, small or large, can tolerate larger relative changes in flow rates?

8.11. Referring to the data of Example 8.1, how long would it take for 1-μm ash particles (equivalent spherical diameter) disgorged to an altitude of 20 km to settle out of the atmosphere down to sea level? How far around the world would particles of this size go in this time?

Data: Average westerly wind speed at 45° latitude (close to the location of the volcano) is 800 km/day. Ignore updrafts and downdrafts, in the long run they should cancel out.

8.12. As part of a new chemical process, small spherical seed particles ($\rho_s = 8500$ kg/m^3, $d_p = 50$ μm) are continuously fed to a hot fluidized bed reactor (629°C) which is fluidized by entering gas. A portion of this gas decomposes in the hot bed, depositing a spongy coating of solids ($\rho_s = 500$ kg/m^3) on the particles which grow in size. The well-mixed solids with their coating leave the reactor. Because the capacity of the reactor depends primarily on the gas throughput, we want to push this up as high as practical. This flow rate is limited, in turn, by elutriation. Estimate the size of particle which will first elutriate from the reactor, and estimate the exiting gas velocity at which this will happen.

Data for hot gas: $(mw) = 0.030$ kg/m^3, $\mu = 2 \times 10^{-5}$ kg/m s, $\pi = 125$ kPa.

Part II   HEAT EXCHANGE

The second part of this volume deals with the exchange of heat from one flowing stream, whether it be solid, liquid, or gas, to another, and the many different kinds of devices, called heat exchangers, which can be used to do this. But before we get to this, we first introduce the three mechanisms of heat transfer and consider their interaction. Chapter 9 thus presents some of the findings on these three mechanisms of heat transfer. Chapter 10 then shows how to treat situations which involve more than one mechanism of heat transfer, and Chapter 11 considers unsteady state heating and cooling of objects. Chapter 12 onward then uses this information for the design of the three major types of heat exchangers: the recuperator, the direct contact exchanger, and the regenerator. Finally, Chapter 16 presents a collection of problems which uses ideas and findings from various chapters in this book.

Chapter 9   THE THREE MECHANISMS OF HEAT TRANSFER: CONDUCTION, CONVECTION, AND RADIATION

In general, heat flows from here to there by three distinct mechanisms:

- by *conduction*, or the transfer of energy from matter to adjacent matter by direct contact, without intermixing or flow of any material.
- by *convection*, or the transfer of energy by the bulk mixing of clumps of material. In *natural convection* it is the difference in density of hot and cold fluid which causes the mixing. In *forced convection* a mechanical agitator or an externally imposed pressure difference (by fan or compressor) causes the mixing.
- by *radiation* such as light, infrared, ultraviolet and radio waves which emanate from a hot body and are absorbed by a cooler body.

In turn, let us briefly summarize the findings on these three mechanisms of heat transfer.

I. HEAT TRANSFER BY CONDUCTION

Conduction refers to the transfer of heat from the hotter to the colder part of a body by direct molecular contact, not by gross movement of clumps of hot material to the cold region. At steady state the rate of heat transfer depends on the nature of the material and the temperature differences, and is expressed by Fourier's law as:

$$\dot{q}_x = -kA\frac{dT}{dx} \qquad (9.1)$$

where $\dot{q}_x$ is the rate of heat transfer in the x direction, [W]; A is the area normal to the direction of heat flow, [m^2]; dT/dx is the temperature gradient in the x direction [K/m]; and k is the thermal conductivity, defined as the heat going through a cube of the material in question 1 m on a side resulting from a temperature difference on opposite faces of 1°C. Table 9.1 gives k values for various materials [W/m K].

The minus sign in this equation tells that heat flows from regions of higher to lower temperature, not the other way round, and shows that the second law of thermodynamics is at work.

Table 9.1. Short Table of Thermal Conductivities
for Materials at Room Temperature[a]

Material	k, W/m K	Material	k, W/m K
Gases		Solids	
SO$_2$	0.009	Styrofoam	0.036
CO$_2$, H$_2$	0.018	Corrugated	
		cardboard	0.064
H$_2$O	0.025	Paper	0.13
Air	0.026	Sand, dry	0.33
Liquids		Glass	0.35–1.3
Gasoline	0.13	Ice	2.2
Ethanol	0.18	Lead	34
Water	0.61	Steel	45
Mercury	8.4	Aluminum	204
Sodium	85	Copper	380

[a]For additional values see Appendices O and U.

The complete equation for steady-state heat conduction in any arbitrary direction through an isotropic material, without heat generation, is

$$\dot{q} = -kA(\nabla T), \qquad k = \text{constant} \tag{9.2}$$

Fourier's equation has been integrated for various simple geometries. Here are some steady state solutions:

A. Flat Plate, Constant k

$$\dot{q}_x = -kA\frac{T_2 - T_1}{x_2 - x_1} = -kA \cdot \frac{T_2 - T_1}{L} \tag{9.3}$$

B. Flat Plate, $k = k_0 (1 + \beta T)$

$$\dot{q}_x = -k_0 A\frac{(T_2 - T_1) + (\beta/2)(T_2^2 - T_1^2)}{x_2 - x_1} = k_{\text{mean}}A(T_2 - T_1) \tag{9.4}$$

where

$$k_{\text{mean}} = \frac{k_1 + k_2}{2}$$

C. Hollow Cylinders, Constant k

$$\dot{q}_r = -2\pi kL\frac{T_2 - T_1}{\ln(r_2/r_1)} \tag{9.5}$$

heat flows outward

D. Hollow Sphere, Constant k

$$\dot{q}_r = -4\pi k r_1 r_2\frac{T_2 - T_1}{r_2 - r_1} \tag{9.6}$$

heat flows outward

E. Series of Plane Walls

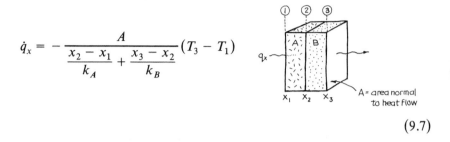

$$\dot{q}_x = -\frac{A}{\dfrac{x_2 - x_1}{k_A} + \dfrac{x_3 - x_2}{k_B}}(T_3 - T_1)$$

A = area normal to heat flow

$$(9.7)$$

F. Concentric Cylinders

$$\dot{q}_r = -\frac{2\pi L}{\dfrac{\ln(r_2/r_1)}{k_A} + \dfrac{\ln(r_3/r_2)}{k_B}}(T_3 - T_1) \qquad (9.8)$$

G. Concentric Spheres

$$\dot{q}_r = -\frac{4\pi}{\dfrac{r_2 - r_1}{k_A r_1 r_2} + \dfrac{r_3 - r_2}{k_B r_2 r_3}}(T_3 - T_1) \qquad (9.9)$$

H. Other Shapes

For nonsimple geometries or for nonuniform temperatures at the boundaries the heat flow can only be obtained by solving Fourier's equation by numerical or graphical methods [see Welty (1978) or McAdams (1954)].

I. Contact Resistance

When heat flows across two touching plane walls, an extra resistance normally is found at the interface because the contacting surfaces are not quite smooth. This results in a sharp temperature drop at the surface. The heat flow across the interface can then be related to the temperature drop

across the interface by

$$q = -h_c A(T_2'' - T_2')$$

where h_c is defined as the contact heat transfer coefficient.

Overall, the heat flow across the two walls will then involve three resistances in series:

Across wall A: $$\dot{q}_x = -k_A A \frac{T_2' - T_1}{x_2 - x_1}$$

Across the interface: $$\dot{q}_x = -h_c A(T_2'' - T_2')$$

Across wall B: $$\dot{q}_x = k_B A \frac{T_3 - T_2'}{x_3 - x_2}$$

Noting that the $\dot{q}$'s are all equal we can combine the above equations to eliminate the intermediate temperatures T_2' and T_2'', ending up with

$$\dot{q}_x = -\frac{1}{\dfrac{x_2 - x_1}{k_A} + \dfrac{1}{h_c} + \dfrac{x_3 - x_2}{k_B}}(T_3 - T_1) \qquad (9.10)$$

Equations analogous to the above can be developed for concentric spheres, concentric cylinders, and other shapes.

II. HEAT TRANSFER BY CONVECTION

When hot fluid moves past a cool surface heat goes to the wall at a rate which depends on the properties of the fluid, and whether it is moving by natural convection, by laminar flow, or by turbulent flow. To account for this form of heat transfer, Prandtl, in 1904, invented the concept of a boundary layer in which all the resistance to heat transfer is located. This idealization led to great simplifications and was enthusiastically adopted by practically all workers [see Adiutori (1974) for a vigorous dissenter].

With this way of viewing things, and with the thickness of boundary layer δ we have

$$\dot{q} = -kA\frac{T_{\text{fluid}} - T_{\text{wall}}}{\delta} = -kA\frac{\Delta T}{\delta}$$

Because δ cannot be estimated independently we combine it with k to give

$$\dot{q} = -(k/\delta)A\Delta T = -hA\Delta T$$

where, by definition,

$$h = \text{heat transfer coefficient, } \left[\text{W}/\text{m}^2\text{ K}\right]$$

Note that h incorporates the thickness of an idealized boundary layer which will give the actual heat transfer rate. This quantity h is extremely useful since it is the rate coefficient which allows us to estimate the heat transfer rate in any particular situation.

Values of h have been measured in all sorts of situations, correlated with the properties of the fluid C_p, ρ, μ, k, the flow conditions u, and the system geometry d, and compactly summarized in dimensionless form. The following sampling of correlations come from McAdams (1954) or Perry and Chilton (1973) unless otherwise noted.

A. Turbulent Flow in Pipes

For both heating and cooling of most normal fluids (Pr = 0.7–700) in fully turbulent flow (Re > 10,000), moderate ΔT, and with physical properties measured at bulk conditions:

$$\underbrace{\frac{hd}{k}}_{\substack{\text{Nusselt} \\ \text{number}}} = 0.023 \underbrace{\left[1 + \left(\frac{d}{L}\right)^{0.7}\right]}_{\substack{\text{Entrance} \\ \text{effect}}} \underbrace{\left[1 + 3.5\frac{d}{d_{\text{coil}}}\right]}_{\substack{\text{For coiled} \\ \text{pipes}}} \left(\frac{dG}{\mu}\right)^{0.8} \underbrace{\left(\frac{C_p\mu}{k}\right)^{1/3}}_{\substack{\text{Prandtl} \\ \text{number}}} \underbrace{\left(\frac{\mu}{\mu_w}\right)^{0.14}}_{\substack{\text{At wall} \\ \text{temperature}}}$$

($= u\rho$, where u = mean velocity)

$$(9.11)$$

A simplified approximation for common gases (error $\pm 25\%$):

$$h = 0.0018 \frac{C_p G^{0.8}}{d^{0.2}} \quad [\text{W/m}^2\,\text{K}] \tag{9.12}$$

and a simplified approximation for cooling and heating of water:

$$h = 91(T + 68)\frac{u^{0.8}}{d^{0.2}} \quad [\text{W/m}^2\,\text{K}] \quad \text{with } T \text{ in } °C \tag{9.13}$$

B. Turbulent Flow in Noncircular Conduits

1. Rectangular cross section. Use the equation for circular pipes, equation (9.11), with the following two modifications:

$$h_{\text{rect}} \cong 0.76\, h_{\text{pipes}} \tag{9.14}$$

and replace the pipe diameter with an equivalent diameter defined as:

$$d_e = 4\left(\begin{array}{c}\text{hydraulic}\\\text{radius}\end{array}\right) = 4\left(\frac{\text{cross-sectional}}{\text{perimeter}}\right) = \frac{2d_1 d_2}{d_1 + d_2} \tag{9.15}$$

2. Annular passage. For heat flow to the inner tube wall

$$\frac{h_i d_e}{k} = 0.20\left(\frac{d_e G}{\mu}\right)^{0.8}\left(\frac{C_p \mu}{k}\right)^{1/3}\left(\frac{d_o}{d_i}\right)^{0.53} \tag{9.16}$$

where

$$d_e = 4\left(\begin{array}{c}\text{hydraulic}\\\text{radius}\end{array}\right) = d_o - d_i \tag{9.17}$$

To the outer tube wall use equation (9.11) for circular pipes, but with the pipe diameter replaced by d_e of equation (9.17).

C. Transition Regime in Flow in Pipes

In the transition regime, $2100 < \text{Re} < 10,000$:

$$\frac{hd}{k} = 0.116\left[\left(\frac{dG}{\mu}\right) - 125\right]\left(\frac{C_p\mu}{k}\right)^{1/3}\left[1 + \left(\frac{d}{L}\right)^{2/3}\right]\left(\frac{\mu}{\mu_w}\right)^{0.14} \quad (9.18)$$

D. Laminar Flow in Pipes (McAdams, Chap. 9)

For heating and cooling in both vertical and horizontal pipes, $\text{Re} < 2100$

$$\frac{hd}{k} = 1.86\underbrace{\left[\frac{dG}{\mu} \cdot \frac{C_p\mu}{k} \cdot \frac{d}{L}\right]^{1/3}}_{\text{Graetz number}}\left(\frac{\mu}{\mu_w}\right)^{0.14} \quad (9.19)$$

Perry and Chilton (1973) give numerous other expressions.

E. Laminar Flow in Pipes, Constant Heat Input Rate at the Wall (Kays and Crawford, 1980)

When the velocity and temperature profiles are fully developed (away from the entrance region), axial dispersion theory predicts that

$$hd/k = 4.36 \quad (9.20)$$

In evaluating h the term ΔT is defined as the difference in temperature between the wall at position x and the mixing cup temperature of the flowing fluid at the same position. This situation is found when using electrical resistance heating or radiant heating.

Theory shows that the laminar velocity profile is fully developed at about

$$x/d = 0.05 \text{ Re} \quad (9.21)$$

and that the thermal profile is fully developed at about

$$x/d = 0.05 \text{ Re} \cdot \text{Pr} \quad (9.22)$$

Thus, equation (9.20) only applies in tubes much longer than the larger of equations (9.21) and (9.22). Let us look at a few typical entry lengths at Re = 100:

		Entry length, x/d		
Fluid	Pr	from velocity profile	from temperature profile	Slower developing profile
Liquid metal	0.01	5	0.05	Velocity
Water	1	5	5	Same
Oil	100	5	500	Temperature

These values show that for liquid metals or ordinary aqueous fluids the entry length is rather short. However if oil or some other high Prandtl number fluid is flowing through the pipe, then the entry length may become substantial, and the value of h predicted by equation (9.20) will be too low. See Kays and Crawford, pg. 114 (1980) for h values for short pipes, and see Perry and Chilton (1973) for h values in other shaped ducts.

F. Laminar Flow in Pipes, Constant Wall Temperature (Kays and Crawford, 1980)

This situation is approached when a process with high h occurs on the outside of the tubes (boiling, condensation, transfer to finned tubes). Here theory says that in the region of fully developed laminar velocity and temperature profiles

$$\frac{hd}{k} = 3.66 \qquad (9.23)$$

Again, this equation only applies when the pipe is much longer than the entry length given by equations (9.17) and (9.18). For shorter pipes the h

value predicted by equation (9.23) will be too low. Kays and Crawford, pg. 128 (1980), give h values for short pipes, and Perry and Chilton (1973) give h values for other shaped ducts.

G. Flow of Gases Normal to a Single Cylinder

Over a very wide range of Reynolds numbers experimental results can be correlated by

$$\frac{hd}{k_f} = A \left(\frac{dG}{\mu_f}\right)^n \left(\frac{C_p\mu}{k_f}\right)^{0.3} \tag{9.24}$$

where subscript f refers to properties of the gas at the film temperature estimated as

$$T_f = \frac{T_{\text{bulk fluid}} + T_{\text{wall}}}{2}$$

and where the constants A and n are given in Table 9.2. For air at 93°C and Re = 1000–50,000 we have the following simplified equation:

$$h = 0.0018 \frac{C_p G^{0.6}}{d^{0.4}} \quad \left[\text{W}/\text{m}^2 \text{ K}\right] \tag{9.25}$$

H. Flow of Liquids Normal to a Single Cylinder

For Re = 0.1–300 the data are correlated by

$$\frac{hd}{k_f} = \left[0.35 + 0.56\left(\frac{dG}{\mu_f}\right)^{0.52}\right]\left(\frac{C_p\mu}{k_f}\right)^{0.3} \tag{9.26}$$

Table 9.2. Constants in Equation (9.24)
for Flow Normal to Single Cylinders

$\dfrac{du\rho_f}{\mu_f}$	A	n	$\dfrac{hd}{k_f}$ for air, from equation (9.20)
1–4	0.960	0.330	0.890–1.42
4–40	0.885	0.385	1.40–3.40
40–4000	0.663	0.466	3.43–29.6
4000–40,000	0.174	0.618	29.5–121.
40,000–250,000	0.257	0.805	121.–528.

I. Flow of Gases Past a Sphere

$$\frac{hd}{k_f} = 2 + 0.6\left(\frac{dG}{\mu_f}\right)^{0.5}\left(\frac{C_p\mu}{k}\right)^{1/3}_f \qquad \text{for } \left(\frac{dG}{\mu_f}\right) < 325 \qquad (9.27)$$

$$\frac{hd}{k_f} = 0.4\left(\frac{dG}{\mu_f}\right)^{0.6}\left(\frac{C_p\mu}{k}\right)^{1/3}_f \qquad \text{for } \left(\frac{dG}{\mu_f}\right) = 325\text{–}70{,}000 \qquad (9.28)$$

J. Flow of Liquids Past a Sphere

$$\frac{hd}{k_f} = \left[0.97 + 0.68\left(\frac{dG}{\mu_f}\right)^{0.52}\right]\left(\frac{C_p\mu}{k}\right)^{0.3}_f \qquad (9.29)$$

K. Other Geometries

For tube banks, h values can be up to 50% higher than for single tubes, the actual value depending on the number of rows and the geometry used. For tube banks, coiled tubes, tubes of noncircular cross section, finned tubes, and many other situations, see McAdams (1954), Chap. 10.

L. Condensation on Vertical Tubes

The theoretical equation derived by Nusselt in 1916 is still recommended today

$$\frac{hL}{k_l} = 0.925\left(\frac{L^3\rho_l^2 g}{\mu_l\Gamma}\right)^{1/3} = 0.943\left(\frac{L^2\rho_l^2 g\lambda}{k_l\mu_l\Delta T}\right)^{1/4} \qquad (9.30)$$

where

$$\Gamma = \left(\frac{\text{flow rate of condensate from the tube}}{\text{circumference}}\right) = \frac{F_l}{\pi d} \qquad [\text{kg/s m}] \qquad (9.31)$$

For steam condensing at atmospheric conditions this equation reduces to

$$h = 0.97\left(\frac{d}{F_l}\right)^{1/3} \quad [\text{W}/\text{m}^2 \text{ K}] \quad (9.32)$$

M. Agitated Vessels to Jacketed Walls

For various types of agitators we have the general expression

$$\underbrace{\frac{hd_{\text{jacket}}}{k} = a\left(\underbrace{\frac{L_p^2 N_r \rho}{\mu}}_{\substack{\text{Reynolds number for} \\ \text{agitated vessels}}}\right)^{b}\left(\frac{C_p\mu}{k}\right)^{1/3}\left(\frac{\mu}{\mu_w}\right)^{m}} \quad (9.33)$$

where the constants a, b, and m are given in Table 9.3.

N. Single Particles Falling through Gases and Liquids (Ranz and Marshall, 1952)

$$\frac{hd_p}{k} = 2 + 0.6\left(\frac{d_p u\rho}{\mu}\right)^{1/2}\left(\frac{C_p\mu}{k}\right)^{1/3} \quad (9.34)$$

Table 9.3. Constants in Equation (9.33) for Heat Transfer
to the Walls of Agitated Vessels

Type of agitator	a	b	m	Range of Re
Paddle	0.36	2/3	0.21	$300\text{–}3 \times 10^5$
Pitched blade turbine	0.53	2/3	0.24	80–200
Disk, flat blade turbine	0.54	2/3	0.14	$40\text{–}3 \times 10^5$
Propeller	0.54	2/3	0.14	2×10^3
Anchor	1.0	1/2	0.18	10–300
Anchor	0.36	2/3	0.18	300–40000
Helical ribbon	0.633	1/2	0.18	$8\text{–}10^5$

O. Fluid to Particles in Fixed Beds (Kunii and Levenspiel, 1979)

(a) For beds of fine solids

with gases: $\quad \dfrac{hd_p}{k} = 0.012\ \mathrm{Re}_p^{1.6}\mathrm{Pr}^{1/3} \qquad$ for $\mathrm{Re}_p < 100 \quad$ (9.35)

with liquids: $\quad \dfrac{hd_p}{k} = 0.16\ \mathrm{Re}_p^{1.6}\mathrm{Pr}^{1/3} \qquad$ for $\mathrm{Re}_p < 10 \qquad$ (9.36)

(b) For coarse solids with both gases and liquids

$$\frac{hd_p}{k} = 2 + 1.8\ \mathrm{Re}_p^{1/2}\mathrm{Pr}^{1/3} \quad \begin{cases} \text{for } \mathrm{Re}_p > 100,\ \text{gases} \\ \text{for } \mathrm{Re}_p > 10,\ \text{liquids} \end{cases} \qquad (9.37)$$

where $\mathrm{Re}_p = (d_p u_0 \rho / \mu)$ and $u_0 =$ superficial velocity (upstream velocity, or in vessel with no solids).

P. Gas to Fluidized Particles

The heat transfer coefficient is difficult to measure in this situation, so until reliable data becomes available the following equation is suggested as a conservative estimate of h.

$$\frac{hd_p}{k} = 2 + 0.6\left(\frac{d_p u_0 \rho}{\mu}\right)^{1/2}\left(\frac{C_p \mu}{k}\right)^{1/3} \qquad (9.38)$$

Q. Fluidized Beds to Immersed Tubes

For beds of fine particles, or $\mathrm{Re}_{mf} < 12.5$, Botterill (1983) recommends the following simple dimensional expression (in SI units):

$$\frac{hd_p}{k_g} = 25\frac{d_p^{0.64}\rho_s^{0.2}}{\left[k_g\ (\text{at bed temperature})\right]^{0.4}} \qquad (9.39)$$

For beds of large particles, or $\mathrm{Re}_{mf} > 12.5$, Botterill suggests using

$$\frac{hd_p}{k_g} = 0.7\left[d_p^{0.5}\mathrm{Ar}^{0.39} + \mathrm{Ar}^{0.15}\right] \qquad (9.40)$$

where Ar is the Archimedes number (see Appendix T).

R. Natural Convection

Slow-moving fluids passing by hot surfaces give larger than expected h values. This is because of natural convection. The particular variables which

characterize natural convection are combined into a dimensionless group, the Grashof number, defined as

Correlations for natural convection are often of the form:

$$\text{Nu} = A[\text{Gr} \cdot \text{Pr}]^B$$

or

$$\frac{hL}{k} = A\left[\left(\frac{L^3\rho^2 g\beta\Delta T}{\mu_f^2}\right)\left(\frac{C_p\mu_f}{k_f}\right)\right]^B \qquad (9.41)$$

or

$$Y = AX^B$$

S. Natural Convection—Vertical Plates and Cylinders, $L > 1$ m

Laminar: $Y = 1.36\ X^{1/5}$ for $X < 10^4$ (9.42)

Laminar: $Y = 0.55\ X^{1/4}$ for $X = 10^4\text{--}10^9$ (9.43)

Turbulent: $Y = 0.13\ X^{1/3}$ for $X > 10^9$ (9.44)

Simplified equations for air at room conditions:

$$h = 1.4\left(\frac{\Delta T}{L}\right)^{1/4} \quad [\text{W/m}^2\ \text{K}] \quad \text{for laminar regime} \quad (9.45)$$

$$h = 1.3(\Delta T)^{1/3} \quad [\text{W/m}^2\ \text{K}] \quad \text{for turbulent regime} \quad (9.46)$$

and for water at room conditions:

$$h = 120(\Delta T)^{1/3} \quad [\text{W/m}^2\ \text{K}] \quad \text{for } X > 10^9 \quad (9.47)$$

T. Natural Convection—Spheres and Horizontal Cylinders, $d < 0.2$ m

Laminar: $Y = 0.53\ X^{1/4}$ for $X = 10^3\text{–}10^9$ $\qquad$ (9.48)

Turbulent: $Y = 0.13\ X^{1/3}$ for $X > 10^9$ $\qquad$ (9.49)

For $X < 10^4$, see Perry and Chilton (1973).
Simplified equations for air at room conditions:

$$h = 1.3(\Delta T/L)^{1/4} \quad [\text{W/m}^2\ \text{K}] \quad \text{for laminar regime} \quad (9.50)$$

$$h = 1.2(\Delta T)^{1/3} \quad [\text{W/m}^2\ \text{K}] \quad \begin{array}{l}\text{for turbulent regime,}\\ \text{the usual case for pipes}\end{array} \quad (9.51)$$

U. Natural Convection for Fluids in Laminar Flow Inside Pipes

In laminar flow, when $\text{Gr} > 1000$, natural convection sets up an appreciable secondary flow of fluid in the pipe which in turn increases the heat transfer coefficient. In this situation equations (9.19), (9.20), and (9.23) for laminar flow should include the additional multiplying factor:

$$0.87(1 + 0.015 \, \text{Gr}^{1/3}) \tag{9.52}$$

For turbulent flow no such correction is needed because the tendency to set up a secondary flow pattern is effectively overwhelmed by the vigorous turbulent eddies.

V. Natural Convection—Horizontal Plates

(a) For heated plates facing up, or cooled plates facing down:

Laminar: $Y = 0.54 \, X^{1/4}$ for $X = 10^5 - 2 \times 10^7$ (9.53)

Turbulent: $Y = 0.14 \, X^{1/3}$ for $X = 2 \times 10^7 - 3 \times 10^{10}$ (9.54)

(b) For heated plates facing down, or cooled plates facing up:

Laminar: $Y = 0.27 \, X^{1/4}$ for $X = 3 \times 10^5 - 3 \times 10^{10}$ (9.55)

(c) The three corresponding simplified equations for air at room conditions:

$$h = 1.3\left(\frac{\Delta T}{L}\right)^{1/4} \quad [\text{W/m}^2 \, \text{K}] \quad \text{for laminar regime} \tag{9.56}$$

$$h = 1.5\Delta T^{1/3} \quad [\text{W/m}^2 \, \text{K}] \quad \text{for turbulent regime} \tag{9.57}$$

$$h = 0.64\left(\frac{\Delta T}{L}\right)^{1/4} \quad [\text{W/m}^2 \, \text{K}] \quad \text{for laminar regime} \tag{9.58}$$

W. Other Situations

Heat transfer coefficients for boiling, condensation, high-velocity gas flow (compressibility effects and supersonic flow), high-vacuum flow, and many other situations, have been studied and reported in the vast heat transfer literature, and are well condensed in McAdams (1954), in Perry and Chilton (1973), and in Cavaseno (1979).

III. HEAT TRANSFER BY RADIATION

All materials emit, absorb, and transmit radiation to an extent which is strongly dependent on their temperature. Let

$$\alpha_{1 \leftarrow 2} = \frac{\text{energy absorbed by a surface at } T_1}{\text{energy incident coming from a source } T_2}, \qquad \text{the absorptivity}$$

(9.59)

The absorptivity varies from 0 to 1. The perfect absorber has $\alpha = 1$ and is called a *blackbody*. Next let

$$\varepsilon_1 = \frac{\text{energy emitted by a surface at } T_1}{\text{energy emitted by an ideal emitter, a blackbody, at } T_1}, \qquad \text{emissivity}$$

(9.60)

and

$$\tau_1 = \frac{\text{energy transmitted through the body at } T_1}{\text{energy incident}}, \qquad \text{transmittance}$$

(9.61)

Then the fraction of energy reflected is $1 - \alpha - \tau$.

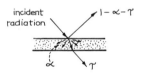

A. Radiation from a Body

The energy emitted from surface A_1 of a body is strongly dependent on the temperature and nature of the surface, and is given by

$$\dot{q}_{1 \rightarrow} = \sigma A_1 \varepsilon_1 T_1^4 \quad [\text{W}] \tag{9.62}$$

where the radiation constant

$$\sigma = 5.67 \times 10^{-8} \ \text{W/m}^2 \ \text{K}^4 \tag{9.63}$$

is called the Stefan–Boltzmann constant. Equation (9.62) is called the Stefan–Boltzmann law of radiation, and the fourth power of temperature is a consequence of the second law of thermodynamics.

B. Radiation onto a Body

The energy absorbed by a surface A_1 which is at T_1 from blackbody surroundings at T_2 is given by

$$\dot{q}_{1 \leftarrow 2} = \sigma A_1 \alpha_{1 \leftarrow 2} T_2^4 \quad [\text{W}]$$

surroundings
or enclosure
at T_2

surface A_1
at T_1, with α_1

T_2
A_2

$$(9.64)$$

C. Energy Interchange between a Body and Its Enveloping Surroundings

The energy interchange between a surface A_1 at T_1 and any kind of surroundings at T_2, from equations (9.62) and (9.64), is then

$$\dot{q}_{12} = \sigma A_1 \left[\varepsilon_1 T_1^4 - \alpha_{1 \leftarrow 2} T_2^4 \right]$$

A_1, T_1, ε_1
$\alpha_{1 \leftarrow 2}$

enveloping
surroundings
A_2, T_2

$$(9.65)$$

D. Absorptivity and Emissivity

If an object and its surroundings are both at T_1, then the object does not gain or lose heat. Thus, equation (9.65) becomes

$$\dot{q}_{12} = 0 = \sigma A_1 \left[\varepsilon_1 T_1^4 - \alpha_{1 \leftarrow 1} T_1^4 \right]$$

Now the value of α and of ε can vary greatly with the type of surface and with temperature, as shown in Table 9.4. However, at any particular temperature T_1, the above expression shows that

$$\varepsilon_1 = \alpha_{1 \leftarrow 1} \qquad (9.66)$$

Table 9.4. Short Table of Absorptivities and Emissivities
of Various Materials[a]

Material	For solar radiation (~ 5000 K) onto a surface at room temperature, $\alpha_{room \leftarrow solar}$	For room temperature radiation, $\varepsilon_{room} = \alpha_{room \leftarrow room}$
Ag, polished	0.07	0.01
Al, bright foil or polished	0.1–0.3	0.04–0.09
Cu, polished	0.18	0.02–0.04
Galvanized iron, weathered	0.89	0.23–0.28
Hg, clean	—	0.09
Stainless steel #301, polished	0.37	0.16
White paint, gloss	0.18	0.92–0.96
Black paint, flat	0.97	0.96–0.98
Aluminum paint	0.55	0.51–0.67
Asphalt pavement, clean	0.93	—
Concrete, rough	—	0.94
Earth, plowed field	0.75	—
Grass	0.75–0.80	—
Gravel	0.29	—
Red brick, rough	0.7–0.75	0.93
Roofing paper, black	—	0.91
White paper	0.28	0.95
Wood	—	0.90–0.04
Snow, clean	0.2–0.35	0.82
Ice	—	0.97
Water, deep	—	0.96

[a] Taken from references in this chapter.

This means that the absorptivity of a surface for T_1 radiation equals the emissivity of that surface when it is at T_1.

E. Greybodies

An object whose absorptivity is the same for all temperature radiation is called a greybody. So, for a greybody

$$\alpha = \varepsilon = \text{const, at all temperatures}$$

The greybody approximation is often used since it greatly simplifies difficult analyses.

F. Radiation between Two Adjacent Surfaces

If the facing surfaces are close enough so that all radiation leaving one surface hits the other then the heat interchange is

negligible loss from edges

$$\dot{q}_{12} = \frac{\sigma A\left[\alpha_{2 \leftarrow 1}\varepsilon_1 T_1^4 - \alpha_{1 \leftarrow 2}\varepsilon_2 T_2^4\right]}{\alpha_{1 \leftarrow 2} + \alpha_{2 \leftarrow 1} - \alpha_{2 \leftarrow 1}\alpha_{1 \leftarrow 2}} \qquad (9.67)$$

1. For *two facing grey surfaces* $\alpha_{2 \leftarrow 1} = \varepsilon_2$, $\alpha_{1 \leftarrow 2} = \varepsilon_1$, so the above expression reduces to

$$\dot{q}_{12} = \frac{\sigma A\left[T_1^4 - T_2^4\right]}{\dfrac{1}{\varepsilon_1} + \dfrac{1}{\varepsilon_2} - 1} \qquad (9.68)$$

2. For *concentric grey cylinders* we obtain, similarly,

Always the smaller area

$$\dot{q}_{12} = \frac{\sigma A_1\left[T_1^4 - T_2^4\right]}{\dfrac{1}{\varepsilon_1} + \dfrac{A_1}{A_2}\left(\dfrac{1}{\varepsilon_2} - 1\right)} \qquad (9.69)$$

G. Radiation between Nearby Surfaces with Intercepting Shields

If two facing surfaces are separated by a very thin opaque shield, then if $\varepsilon_1 = \varepsilon_2$, while ε_s can be any value, we find

$$T_s^4 = \frac{T_1^4 + T_2^4}{2} \qquad (9.70)$$

from which

$$\dot{q}_{12} = \frac{1}{2}\frac{\sigma A\left[T_1^4 - T_2^4\right]}{\dfrac{1}{\varepsilon_1} + \dfrac{1}{\varepsilon_s} - 1} \qquad (9.71)$$

Thus, a shield of material similar to the two radiating surfaces will halve the radiation transfer between two closely facing surfaces. Extending this analysis, we find for n shields of identical emissivity, that

$$\dot{q}_{12} = \frac{1}{n+1} \frac{\sigma A \left[T_1^4 - T_2^4 \right]}{\dfrac{1}{\varepsilon_1} + \dfrac{1}{\varepsilon_s} - 1} \qquad (9.72)$$

Similarly, for a shield between pipes, or around a sphere,

$$\dot{q}_{12} = \frac{\sigma A_1 \left[T_1^4 - T_2^4 \right]}{\dfrac{1}{\varepsilon_1} + \dfrac{A_1}{A_2}\left(\dfrac{1}{\varepsilon_2} - 1\right) + \dfrac{A_1}{A_s}\left(\dfrac{2}{\varepsilon_s} - 1\right)} \qquad (9.73)$$

In all cases radiation shields reduce the radiative heat interchange between bodies.

H. View Factors for Blackbodies

If both surfaces are black and not close together, then only a portion of the radiation leaving surface 1 is intercepted by surface 2. We call this the

view factor F_{12}, and so the radiation leaving 1 which is intercepted by 2 is

$$\dot{q}_{2 \leftarrow 1} = \sigma A_1 F_{12} T_1^4 \qquad (9.74)$$

Similarly, the radiation leaving 2 which is intercepted by 1 is

$$\dot{q}_{1 \leftarrow 2} = \sigma A_2 F_{21} T_2^4 \qquad (9.75)$$

If both temperatures are equal there can be no net transfer of heat between 1 and 2. Thus we find

$$A_1 F_{12} = A_1 F_{21} \qquad (9.76)$$

The net interchange of heat between these two surfaces is then

$$\dot{q}_{12} = \sigma A_1 F_{12}\left[T_1^4 - T_2^4\right] \quad \text{with } A_1 F_{12} = A_2 F_{21} \qquad (9.77)$$

If the surfaces are grey, not black, the heat exchange is approximated by

$$\dot{q}_{12} = \sigma A F_{12}'\left[T_1^4 - T_2^4\right] \qquad (9.78)$$

where

$$F_{12}' = \cfrac{1}{\cfrac{1}{F_{12}} + \left(\cfrac{1}{\varepsilon_1} - 1\right) + \cfrac{A_1}{A_2}\left(\cfrac{1}{\varepsilon_2} - 1\right)} \qquad (9.79)$$

I. View Factor for Two Black- (or Grey-) Bodies Plus Reradiating Surfaces

The net heat interchange between black surfaces 1 and 2 in the presence of adiabatic reradiating surfaces is given by

$$\dot{q}_{12} = \sigma A_1 \overline{F}_{12}\left(T_1^4 - T_2^4\right) \qquad (9.80)$$

where $\overline{F}_{12}$ depends on F_{12} and the geometry of the reradiating surfaces. After making simplifying assumptions that surfaces 1 and 2 cannot see themselves, and that the reradiating surfaces are all at one temperature, we find that

$$\overline{F}_{12} = \frac{A_2 - A_1 F_{12}^2}{A_1 + A_2 - 2A_1 F_{12}} \qquad (9.81)$$

Because these reradiating surfaces return some of the radiation which would otherwise be lost suggests that $\overline{F}_{12}$ is always larger than F_{12}, and this is so.

If surfaces 1 and 2 are grey then

$$\dot{q}_{12} = \sigma A_1 \mathscr{F}_{12}\left(T_1^4 - T_2^4\right) \qquad (9.82)$$

where $\mathscr{F}_{12}$ is the greybody view factor for systems with reradiating surfaces, and is approximated by

$$\mathscr{F}_{12} = \frac{1}{\dfrac{1}{\overline{F}_{12}} + \left(\dfrac{1}{\varepsilon_1} - 1\right) + \dfrac{A_1}{A_2}\left(\dfrac{1}{\varepsilon_2} - 1\right)} \tag{9.83}$$

This is the most general of view factors.

The appendix in Siegel and Howell (1981) refers to view factors for over 200 different kinds of geometries, and gives equations for 38 of these geometries. Figures 9.1–9.5, from Jakob (1957), show the view factors for five simple geometries.

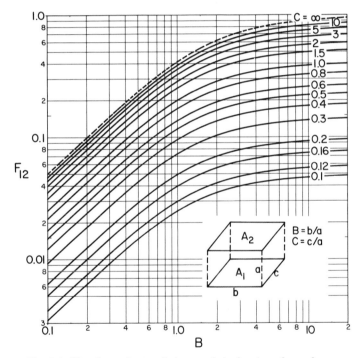

Fig. 9.1. View factor for two facing equal sized rectangular surfaces.

$$F_{12} = \frac{1}{\pi}\left[\frac{1}{BC}\ln\frac{1 + B^2 + C^2 + B^2C^2}{1 + B^2 + C^2} + \frac{2}{C}(1 + C^2)^{1/2}\tan^{-1}\frac{B}{(1 + C^2)^{1/2}}\right.$$

$$\left. - \frac{2}{B}\tan^{-1}C - \frac{2}{C}\tan^{-1}B + \frac{2}{B}(1 + B^2)^{1/2}\tan^{-1}\frac{C}{(1 + B^2)^{1/2}}\right]$$

Fig. 9.2. View factor for two perpendicular rectangular surfaces having a common edge.

$$F_{12} = \frac{1}{\pi B}\left\{\frac{1}{4}\ln\left[(1 + B^2 + C^2)^{B^2 + C^2 - 1}(1 + B^2)^{1 - B^2}\right.\right.$$
$$\left.\times (1 + C^2)^{1 - C^2}(B^2)^{B^2}(C^2)^{C^2}\right]$$
$$- \frac{1}{4}\ln(B^2 + C^2)^{B^2 + C^2} + B\tan^{-1}\frac{1}{B} + C\tan^{-1}\frac{1}{C}$$
$$\left.- (B^2 + C^2)^{1/2}\tan^{-1}\frac{1}{(B^2 + C^2)^{1/2}}\right\}$$

Fig. 9.3. Fraction of radiation leaving a differential surface which is intercepted by a facing rectangular surface, located as shown

$$F_{12} = \frac{1}{2\pi}\left(\frac{B}{(1 + B^2)^{1/2}}\sin^{-1}\frac{C}{(1 + B^2 + C^2)^{1/2}}\right.$$
$$\left.+ \frac{C}{(1 + C^2)^{1/2}}\sin^{-1}\frac{B}{(1 + B^2 + C^2)^{1/2}}\right)$$

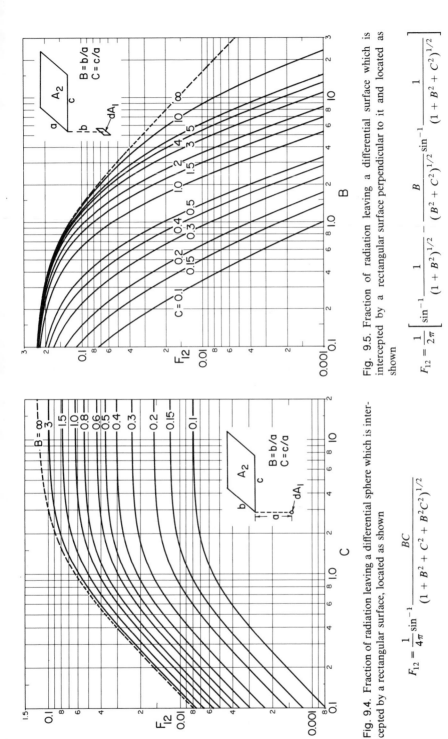

Fig. 9.4. Fraction of radiation leaving a differential sphere which is intercepted by a rectangular surface, located as shown

$$F_{12} = \frac{1}{4\pi} \sin^{-1} \frac{BC}{(1 + B^2 + C^2 + B^2C^2)^{1/2}}$$

Fig. 9.5. Fraction of radiation leaving a differential surface which is intercepted by a rectangular surface perpendicular to it and located as shown

$$F_{12} = \frac{1}{2\pi} \left[\sin^{-1} \frac{1}{(1 + B^2)^{1/2}} - \frac{B}{(B^2 + C^2)^{1/2}} \sin^{-1} \frac{1}{(1 + B^2 + C^2)^{1/2}} \right]$$

J. Extensions

Gases consisting of molecules which are not symmetrical about all three principal axes ($NH_3, CO, CO_2, H_2O, HCl, \ldots$, etc.) absorb and emit significant amounts of radiation at high temperature. Symmetrical molecules ($O_2, N_2, H_2, \ldots$, etc.) do not absorb or emit significantly in the temperature range of practical interest.

Heat interactions between absorbing gases and surfaces are accounted for by a characteristic emissivity and a characteristic view factor, somewhat like two-surface systems. Clouds of fine particles, soot, luminous flames, etc., are treated in the same manner.

We do not take up these subjects here. The reader is referred to the references for further readings on these subjects.

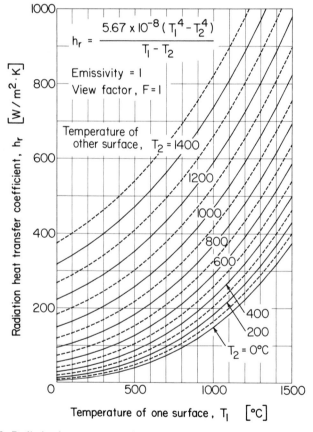

Fig. 9.6. Radiation between two surfaces in terms of a heat transfer coefficient.

K. Estimating the Magnitude of h_r

For design we want to know whether radiative transfer is appreciable compared to the other competing mechanisms of heat transfer and whether it need be considered at all in any analysis. Figure 9.6 is helpful for this purpose. It gives the radiation heat transfer coefficient between two closely facing ($F = 1$) black surfaces ($\varepsilon = \alpha = 1$). To find $h_{r,\text{actual}}$ for a particular situation, lower the h_r value given in this figure to account for $\mathscr{F} < 1$ and $\varepsilon < 1$. Thus

$$h_{r,\text{actual}} = \varepsilon \mathscr{F} h_{r,\text{figure}}$$

If $h_r \ll h_{\text{convection}}$, ignore the radiation contribution to the overall heat transfer. A look at this figure shows that h_r becomes very large at high temperature.

REFERENCES AND RELATED READINGS

Conduction

E. F. Adiutori, *The New Heat Transfer*, Venturo, Cincinnati (1974).
W. H. McAdams, *Heat Transmission*, Third Ed., McGraw-Hill, New York (1954).
J. R. Welty, *Engineering Heat Transfer*, Second Ed., Wiley, New York (1978).

Convection

J. S. M. Botterill, "Fluidized Bed Behavior," in *Fluidized Beds, Combustion and Applications* (J. R. Howard ed.), Applied Science, New York (1983).
V. Cavaseno, ed., *Process Heat Exchange*, pp. 20, 101, 130, 140, McGraw-Hill, New York (1979).
W. M. Kays and M. E. Crawford, *Convective Heat and Mass Transfer*, Second Ed., Chapter 8, McGraw-Hill, New York (1980).
D. Kunii and O. Levenspiel, *Fluidization Engineering*, Krieger Melbourne, FL (1979).
W. H. McAdams, *Heat Transmission*, Third ed., McGraw-Hill, New York (1954).
R. H. Perry and C. H. Chilton, *Chemical Engineers' Handbook*, Fifth Ed., Sec. 10, McGraw-Hill, New York (1973).
W. E. Ranz and W. R. Marshall, Jr., Evaporation from drops, *Chem. Eng. Prog.* **48**, 141 (1952).

Radiation

H. C. Hottel, Radiant heat transmission, *Mech. Eng.* **52**, 699 (1930).

H. C. Hottel and A. F. Sarofim, *Radiative Transfer*, McGraw-Hill, New York (1967).

M. Jakob, *Heat Transfer*, Vol. 2, Wiley, New York (1957).

W. H. McAdams, *Heat Transmission*, Third Ed., Chapter 4, McGraw-Hill, New York (1954).

M. Siegel and J. R. Howell, *Thermal Radiation Heat Transfer*, Second Ed., McGraw-Hill, New York (1981).

Chapter 10   COMBINATION OF HEAT TRANSFER RESISTANCES

Heat loss from a warm room through a wall to the cold outside involves three heat transfer steps in a series: (i) convection at the inside surface of the wall; (ii) conduction through the wall; and (iii) convection on the outside of the wall. Next, consider a fireplace fire. Here heat reaches the room by radiation from the flames and also by convection of moving air. These processes occur in parallel. There are many processes like these which involve a number of heat transfer steps, sometimes in a series, sometimes in parallel, sometimes in a more involved way.

To find the overall effect of a number of heat transfer steps in a series and in parallel we draw on the analogy to electrical theory. For processes in a series the resistances are additive, thus

$$R_{\text{overall}} = R_1 + R_2 + R_3 \qquad (10.1a)$$

or, in terms of conductances

$$\frac{1}{C_{\text{overall}}} = \frac{1}{C_1} + \frac{1}{C_2} + \frac{1}{C_3} \qquad (10.1b)$$

For processes in parallel, it is the conductances which are additive. Thus

$$C_{overall} = C_1 + C_2 + C_3 \qquad (10.2a)$$

or, in terms of resistances,

$$\frac{1}{R_{overall}} = \frac{1}{R_1} + \frac{1}{R_2} + \frac{1}{R_3} \qquad (10.2b)$$

In the series–parallel situations sketched in Fig. 10.1a we have

$$\frac{1}{C_{overall}} = \frac{1}{C_1} + \frac{1}{C_2 + C_3} \qquad (10.3)$$

and for the sketch of Fig. 10.1b

$$C_{overall} = C_1 + \frac{1}{\dfrac{1}{C_2} + \dfrac{1}{C_3}} \qquad (10.4)$$

For processes in series a glance at equation (10.1) shows that the step with the largest resistance dominates and determines, in most part, the overall resistance. Resistances much smaller than this can be ignored.

For processes in parallel matters are quite different, for equation (10.2) shows that the term with largest conductance (hence smallest resistance) dominates and for the most part determines the overall conductance of the process. Conductances much smaller than this can be ignored.

Whenever a particular step in the overall process dominates to the exclusion of all other steps it is called the rate controlling step of the process.

In heat transfer, the individual conductances are represented by the convection coefficient h, the conductivity per unit length $k/\Delta x$, and the radiation coefficient h_r, while the overall conductance for the process is represented by the overall heat transfer coefficient, U. It is this overall

Fig. 10.1. Two series–parallel arrangements of resistances to heat transfer.

transfer coefficient which is of prime importance in heat exchanger design. We now show the form of the overall transfer coefficient U for a few representative situations. The problems at the end of the chapter present other situations.

A. Fluid – Fluid Heat Transfer through a Wall

As shown in Fig. 10.2 the wall material and both liquid films contributes resistance to heat transfer. Thus we have

$$\dot{q} = h_{12}A(T_1 - T_2)$$

$$\dot{q} = \frac{k_{23}A}{\Delta x_{23}}(T_2 - T_3)$$

$$\dot{q} = h_{34}A(T_3 - T_4)$$

Combining and eliminating the intermediate temperatures T_2 and T_3 gives

$$\dot{q} = -UA\Delta T \quad \text{where} \quad \frac{1}{U} = \frac{1}{h_{12}} + \frac{\Delta x_{23}}{k_{23}} + \frac{1}{h_{34}} \quad (10.5)$$

More generally, if there are scale deposits on the surfaces of the separating wall these deposits represent two more resistances in the series, as

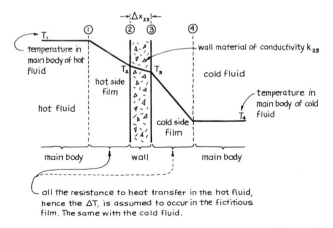

all the resistance to heat transfer in the hot fluid, hence the ΔT, is assumed to occur in the fictitious film. The same with the cold fluid.

Fig. 10.2. Heat transfer from fluid to fluid that are separated by a wall.

Fig. 10.3. Heat transfer across a flat wall which has scale deposits.

shown in Fig. 10.3. In this case we have

$$\frac{1}{U} = \frac{1}{h_2} + \frac{\Delta x_{23}}{k_{23}} + \frac{\Delta x_{34}}{k_{34}} + \frac{\Delta x_{45}}{k_{45}} + \frac{1}{h_{56}} \qquad (10.6)$$

Note that equations (10.5) and (10.6) are of the form of equation (10.1) which represents resistances in a series. Thus, if any particular resistance step is very much larger than all the others (low h or low $k/\Delta x$), it will dominate and determine the overall resistance of the process.

B. Fluid – Fluid Transfer through a Cylindrical Pipe Wall

Consider heat transfer from hot fluid at T_1 to cold fluid at T_6 across a pipe with thin scale coatings (thus $A_2 \cong A_3$ and $A_4 \cong A_5$) as shown in Fig.

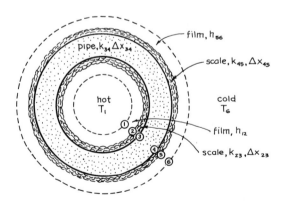

Fig. 10.4. Heat transfer across a scale-coated pipe wall.

10.4. Noting that all the resistances occur in a series we can show that

$$q = -UA\Delta T_{\text{overall}} = U_3 A_3 (T_1 - T_6) = U_4 A_4 (T_1 - T_6)$$

with

Scale   Wall   Scale

$$\frac{1}{U_3 A_3} = \frac{1}{U_4 A_4} = \frac{1}{h_{12} A_3} + \frac{\Delta x_{23}}{k_{23} A_3} + \frac{\Delta x_{34}}{k_{34} A_{34,lm}} + \frac{\Delta x_{45}}{k_{45} A_4} + \frac{1}{h_{56} A_4}$$

$$(10.7)$$

where

$$A_{34,lm} = \frac{A_4 - A_3}{\ln \dfrac{A_4}{A_3}} \cong \frac{A_4 + A_3}{2} \qquad \text{if } \frac{A_4}{A_3} < 2$$

Note that the value of U obtained depends on the area basis chosen and whether it is the inside area or the outside area of the pipe.

C. Conduction across a Wall Followed by Convection and Radiation

Heat transfer by conduction or convection is proportional to the temperature difference of the two locations while radiation transfer is proportional to the difference of T^4. Thus, the latter is very much more temperature sensitive and should dominate at high temperatures, but be negligible at low temperatures.

To be able to assess the relative contributions of the various mechanisms and to be able to combine them, we need to express them in the same measure. We can cast the radiation transfer in terms of a heat transfer coefficient if we put

$$q = h_r A_1 (T_1 - T_2) = \sigma A_1 \mathscr{F}_{12} (T_1^4 - T_2^4)$$

Or the appropriate expression
from Chapter 9

Thus, the radiation heat transfer coefficient is:

$$h_r = \frac{\sigma \mathscr{F}_{12} (T_1^4 - T_2^4)}{T_1 - T_2} \qquad (10.8)$$

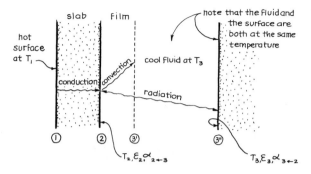

Fig. 10.5. Conduction through a wall followed by convection and radiation in parallel.

Now consider the conduction of heat through a slab followed by convection to a fluid at T_3 and radiation to a facing surface at T_3, as sketched in Fig. 10.5. This is a series–parallel situation as shown in Equation (10.3), or

$$\frac{1}{C_{\text{total}}} = \frac{1}{C_{12}} + \frac{1}{C_{23,\text{conv}} + C_{23,\text{rad}}}$$

In heat transfer language this combination of conductances gives

$$\dot{q}_{13} = -UA\Delta T = UA(T_1 - T_3)$$

where

$$\frac{1}{U} = \frac{\Delta x_{12}}{k_{12}} + \frac{1}{h_{23,\text{conv}} + h_{23,\text{rad}}}$$

$$(10.9)$$

and where, from equation (9.67) for parallel facing source and sink

$$h_{23,\text{rad}} = \frac{\sigma\left[\alpha_{3\leftarrow2}\varepsilon_2 T_2^4 - \alpha_{2\leftarrow3}\varepsilon_3 T_3^4\right]}{(\alpha_{2\leftarrow3} + \alpha_{3\leftarrow2} - \alpha_{3\leftarrow2}\alpha_{2\leftarrow3})(T_2 - T_3)}$$

D. Convection and Radiation to Two Different Temperature Sinks

Now consider a more complex case where heat is lost from surface 1 by convection to fluid at T_2, but also by radiation through the transparent fluid to a parallel surface at T_3, as shown in Fig. 10.6. The heat lost by surface 1

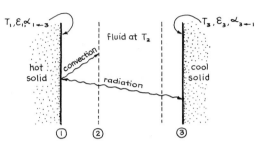

Fig. 10.6. Heat transfer to two
different temperature sinks.

is given by

$$\dot{q}_1 = \dot{q}_{12,\text{conv}} + \dot{q}_{13,\text{rad}}$$

$$= h_{12,\text{conv}} A(T_1 - T_2) + h_{13,\text{rad}} A(T_1 - T_3)$$

where, from equation (9.67),

$$h_{13,\text{rad}} = \frac{\sigma A\left[\alpha_{3 \leftarrow 1}\varepsilon_1 T_1^4 - \alpha_{1 \leftarrow 3}\varepsilon_3 T_3^4\right]}{(\alpha_{1 \leftarrow 3} + \alpha_{3 \leftarrow 1} - \alpha_{1 \leftarrow 3}\alpha_{3 \leftarrow 1})(T_1 - T_3)}$$

In terms of $T_1 - T_2$ we get

$$\dot{q}_{1 \to} = U_{12} A(T_1 - T_2) \qquad \text{where } U_{12} = h_c + h_r\left(\frac{T_1 - T_3}{T_1 - T_2}\right) \quad (10.10)$$

or, in terms of $T_1 - T_3$,

$$\dot{q}_{1 \to} = U_{13} A(T_1 - T_3) \qquad \text{where } U_{13} = h_c\left(\frac{T_1 - T_2}{T_1 - T_3}\right) + h_r \quad (10.11)$$

Note that the temperature ratio appears in an h term whenever that h term is based on one ΔT, while the U and other h term is based on another ΔT.

E. Determination of Gas Temperature

As a final example of the interaction of these different modes of heat transfer consider the determination of the temperature of a hot gas flowing in a pipe. As shown in the sketch of Fig. 10.7 the thermocouple is protected by a shield, a common practice with corrosive gases. First of all, since no

Fig. 10.7. Finding the temperature of a gas with a shielded thermocouple.

heat is lost by the thermocouple or the gas within the shield

$$T_{tc} = T_{sh}$$

Now make a heat balance for the shield. Heat enters by convection from the hot gas; however, heat leaves by net radiation to the cooler walls. Thus, at steady state,

$$\dot{q}_{\substack{\text{convection} \\ \text{to shield}}} + \dot{q}_{\substack{\text{radiation interchange} \\ \text{between shield and walls}}} = 0$$

or

$$\underset{\substack{\text{For gas flow} \\ \text{outside and normal} \\ \text{to a cylinder}}}{h_{\text{conv}}\left(T_g - T_{sh}\right)} = \underset{\substack{\text{For a completely enclosed greybody;} \\ \text{use equation (9.65)}}}{\sigma\varepsilon_{sh}\left(T_{sh}^4 - T_w^4\right)}$$

and noting that $T_{sh} = T_{tc}$

$$T_g = T_{tc} + \frac{\sigma\varepsilon_{sh}}{h_{\text{conv}}}\left(T_{tc}^4 - T_w^4\right) \tag{10.12}$$

The temperature of the shield (hence of thermocouple) will be somewhere between the gas and wall temperature. High h_{conv} and low ε_{sh} will result in the thermocouple measuring the gas temperature, while low h_{conv} and high ε_{sh} will result in the thermocouple measuring the wall temperature.

This type of radiation correction is very important at high temperatures.

F. Extensions

For other more complicated situations, either

- write out all the individual heat interchange terms and eliminate all intermediate temperatures. This procedure is illustrated in section A, above; or

- develop the electrical analog, find the overall conductance or overall resistance, and then replace with heat flow terms. This procedure is illustrated in section C, above.

Usually the latter procedure is simpler, but with more than one source and one sink one must be careful in using this approach, as shown in section D, above.

PROBLEMS ON COMBINING RESISTANCES

Tube Furnace. A pipe carrying fluid passes through a large hot furnace. Develop an expression for the heat absorbed by the fluid using the nomenclature and values given below.

10.1. Assume that the furnace gas and furnace walls are at the same temperature, or $T_1 = T_2$.

10.2. Consider the furnace gas and furnace walls to be at different temperatures, or $T_1 \neq T_2$.

10.3. Consider $T_1 \neq T_2$, and include scale resistance on the inside and the outside of the tube. Renumber all items in Fig. 10.P1 from 1 to 7.

10.4. A transparent gas flows in a channel bounded by two close parallel surfaces which are at different temperatures. Develop an expression for the heat interchange between these surfaces, by convection and by radiation. Assume that both surfaces are grey and have the same emissivity ε.

10.5. *Concentric pipes.* Find an expression for the heat loss from hot fluid flowing in a pipe to the surroundings. The hot pipe is protected by a concentric shield pipe.

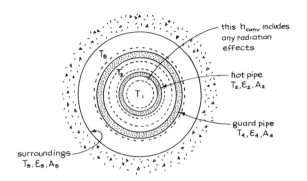

Consider radiation between inner and outer pipe and between outer pipe and surroundings, consider the pipes to have negligible resistance to conduction and to be grey surfaces of emissivity ε, and base the U value on the area of hot pipe A_2. For additional nomenclature see drawing above.

A thermocouple, protected by a stainless-steel shield is inserted in an air preheater duct. At the air velocity flowing past the duct it is estimated that $h_{\text{conv}} = 100 \text{ W/m}^2$ K. Find the temperature of the hot air

10.6. if the thermocouple reads $T_{tc} = 400$ K and if the temperature of the steel walls is $T_w = 300$ K.

10.7. if $T_{tc} = 1000$ K and if $T_w = 900$ K. Note how sharply the error in T_{tc} reading increases (because of the intrusion of radiation) as the temperature level rises.

Chapter 11   UNSTEADY-STATE HEATING AND COOLING OF SOLID OBJECTS

If a hot object is plunged into cold water it cools, but not instantaneously. Two factors govern the cooling rate of the object:

- the film resistance at the surface of the object, characterized by the h value for that situation; and
- the rate of heat flow out of the interior of the object. The governing differential equation for this process is

$$\frac{\partial T_s}{\partial t} = \alpha \left(\frac{\partial^2 T_s}{\partial x^2} + \frac{\partial^2 T_s}{\partial y^2} + \frac{\partial^2 T_s}{\partial z^2} \right) \qquad (11.1)$$

where

$$\alpha = \frac{k_s}{\rho_s C_s}, \qquad \text{thermal diffusivity } \left[\text{m}^2/\text{s} \right]$$

and

$$T_s = \text{temperature at any point in the object} \qquad [\text{K}]$$

A dimensionless measure for conduction, which accounts for both the cooling time and the size of object, is given by the Fourier number

$$\text{Fo} = \frac{\alpha t}{L^2} = \frac{\alpha t}{(V/A)^2} = \frac{k_s}{\rho_s C_s} \cdot \frac{t}{(V/A)^2} \qquad (11.2)$$

where the characteristic length of the object

$$L \begin{cases} = \dfrac{\text{volume}}{\text{surface}} = \dfrac{V}{A} & \text{in general} \\[2ex] = \dfrac{\text{thickness}}{2} & \text{for a slab} \\[2ex] = \dfrac{R}{2} & \text{for a cylinder} \\[2ex] = \dfrac{R}{3} & \text{for a sphere} \end{cases} \qquad (11.3)$$

The relative importance of the surface and the interior resistance terms is measured by the Biot number, a dimensionless group defined as

$$\text{Bi} = \left(\frac{\text{interior resistance}}{\text{surface resistance}} \right) = \frac{h(V/A)}{k_s} = \frac{hL}{k_s} \qquad (11.4)$$

Fig. 11.1. Temperature–time history of a cooling particle for different ranges of Biot numbers.

For a small Biot number the main resistance is in the film, for a large Biot number the main resistance is conduction of heat out of the body. Figure 11.1 shows the temperature–time history in various regimes for a spherical particle.

We first consider the two extreme cases and then the general case where both resistances are important. Design charts present compactly the whole range of situations for various shapes of particles.

I. THE COOLING OF AN OBJECT WHEN ALL THE RESISTANCE IS AT ITS SURFACE (Bi = hL/k_s → 0)

This extreme views the object to be isothermal at any time, the whole object cooling (or heating) with time, as illustrated in Figs. 11.1a and 11.2. This type of analysis where the system in question has uniform properties throughout is called a *lumped parameter analysis*. A heat balance about the hot object being cooled then gives

$$-\dot{q} = \begin{pmatrix} \text{heat transfer rate} \\ \text{through the film} \end{pmatrix} = \begin{pmatrix} \text{rate of heat loss} \\ \text{from the object} \end{pmatrix} \quad [\text{W}]$$

$$= hA(T_s - T_g) = -WC_s \frac{dT_s}{dt} \qquad (11.5)$$
$$\underset{\displaystyle \overset{\displaystyle |}{V\rho_s}}{}$$

Separating and integrating at constant $hA/V\rho_s C_s$ gives

$$\frac{\Delta T}{\Delta T_{\max}} = \frac{T_s - T_g}{T_{s0} - T_g} = e^{-\text{Fo}\cdot\text{Bi}} = e^{-(ht/L\rho_s C_s)} \qquad (11.6)$$

$L = V/A$, characteristic length

at t = 0

T_{s0}

T_g = constant

T_s, at any time

surface area A

uniform temperature throughout

Fig. 11.2

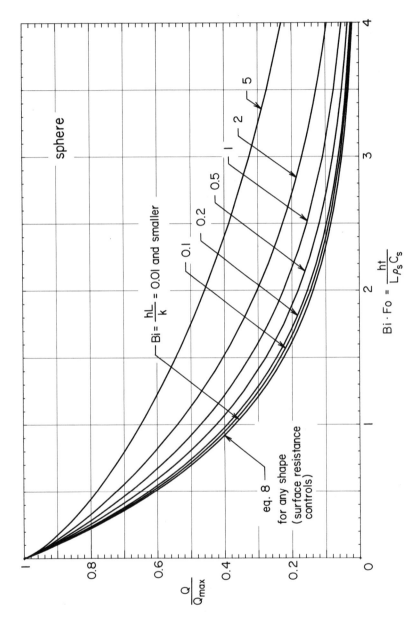

Fig. 11.3. The cooling of any shape of object when surface resistance controls (lowest curve), and the cooling of spheres in general (all the curves).

The instantaneous rate of heat loss from the object is found by combining equations (11.5) and (11.6), or

$$-\dot{q} = -\rho_s C_s V \frac{dT_s}{dt} = hA(T_{s0} - T_g)e^{-Fo \cdot Bi} \tag{11.7}$$

Also, the fractional cooling of the object is found either by integrating equation (11.7), or more simply, by inspection of equation (11.6). Thus

$$\frac{Q}{Q_{max}} = \left(\begin{array}{c} \text{heat left at the} \\ \text{object at time } t \\ \hline \text{total heat which} \\ \text{could be lost} \end{array} \right) = \frac{T_s - T_g}{T_{s0} - T_g} = e^{-Fo \cdot Bi} = e^{-(ht/L\rho_s C_s)}$$

$$\tag{11.8}$$

The lowest curve of Fig. 11.3 represents the equation for this extreme. Note that one curve represents all shapes of solids.

II. THE COOLING OF AN OBJECT HAVING NEGLIGIBLE SURFACE RESISTANCE (Bi = $hL/k_s \to \infty$)

In this extreme, when the hot object is plunged into cold fluid its surface immediately drops to the temperature of the fluid, and conduction within the object is all important. This is illustrated in Fig. 11.1c and in Fig. 11.4. Solving equation (11.1) for a slab gives a rapidly converging infinite series for the temperature T_s at any location and any time

$$\frac{\Delta T}{\Delta T_{max}} = \frac{T_s - T_g}{T_{s0} - T_g} = \frac{4}{\pi}\left(e^{-a}\sin b + \frac{1}{3}e^{-9a}\sin 3b + \frac{1}{5}e^{-25a}\sin 5b + \cdots \right)$$

$$\tag{11.9}$$

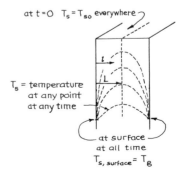

Fig. 11.4. The cooling of an object of negligible surface resistance.

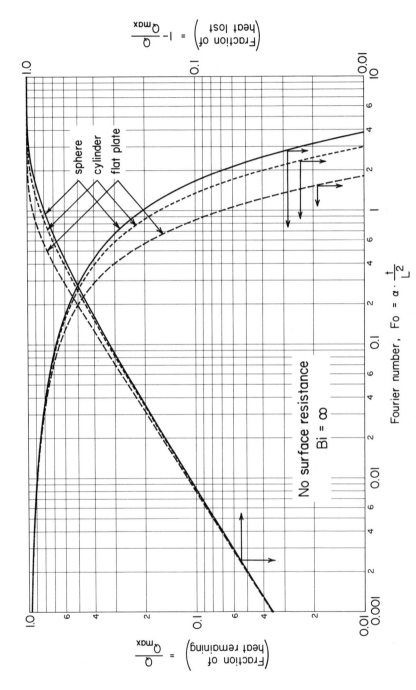

Fig. 11.5. Heat lost and heat remaining in a cooling object for negligible surface resistance. [Prepared by Mator (1982).]

where

$$a = \frac{\pi \alpha t}{4L^2}$$

and

$$b = \frac{\pi l}{2L}$$

The fraction of total heat remaining in the slab is then given by

$$\frac{Q}{Q_{max}} = \left(\begin{array}{c} \text{heat remaining} \\ \text{in slab} \\ \hline \text{total heat initially} \\ \text{in slab} \end{array} \right) = \frac{8}{\pi^2} \left(e^{-a} + \frac{1}{9} e^{-9a} + \frac{1}{25} e^{-25a} + \cdots \right)$$

$$(11.10)$$

Similar equations have been derived for infinite cylinders and for spheres. Figure 11.5 shows both the fraction of heat remaining and the fraction of heat lost from these regular solids. By interpolation between these curves one can estimate the extent of heating or cooling of any irregular solid.

III. THE COOLING OF AN OBJECT WHERE BOTH SURFACE AND INTERNAL RESISTANCE TO HEAT FLOW ARE IMPORTANT

Here (see Fig. 11.1b) we have the conduction equations of the just treated case B with the boundary condition at any time

$$\left(\begin{array}{c} \text{Rate of heat flow out} \\ \text{from surface} \end{array} \right) = -\dot{q} = hA \left(T_{s, l=L} - T_g \right) = k_s A \left(\frac{\partial T_s}{\partial l} \right)_{l=L}$$

$$(11.11)$$

The solutions to these equations have been derived for a number of shapes, and are available in many heat transfer texts [for example, see Gröber, *et al.* (1961) and Boelter *et al.* (1956)]. In all cases these solutions involve slowly converging infinite series, which are tedious to evaluate. However, convenient graphical representations of these solutions have been prepared and are reproduced in Figs. 11.6–11.10 in terms of the following dimensionless

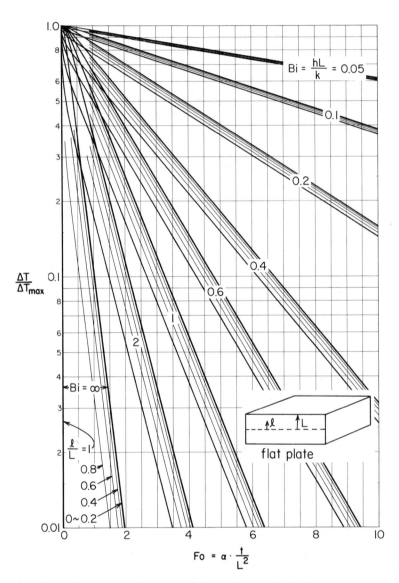

Fig. 11.6a. Temperature distribution within cooling infinite flat plates, general case.

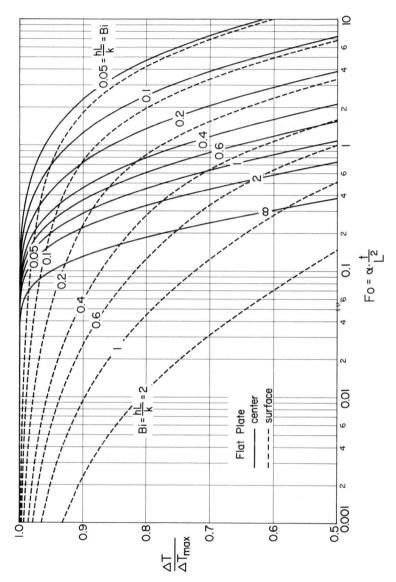

Fig. 11.6b. Top left-hand corner of Fig. 11.6a.

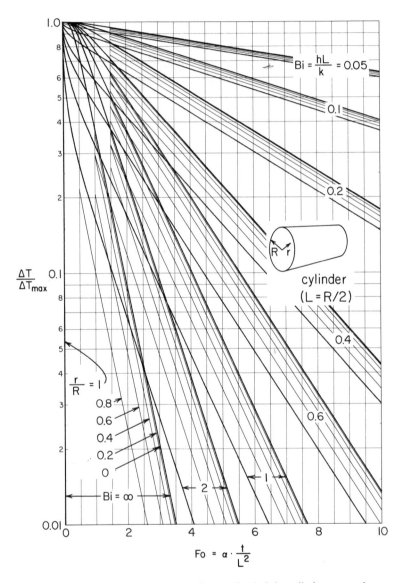

Fig. 11.7a. Temperature distribution within cooling infinite cylinders, general case.

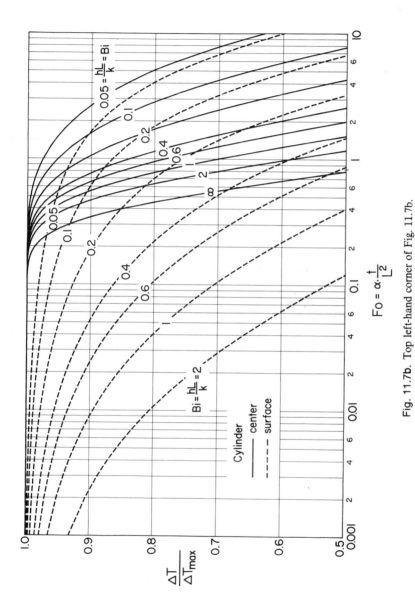

Fig. 11.7b. Top left-hand corner of Fig. 11.7b.

Fig. 11.8a. Temperature distribution within cooling spheres, general case (also see Fig. 11.8b).

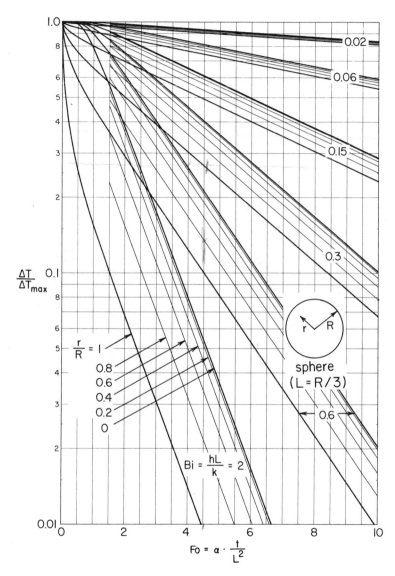

Fig. 11.8b. Temperature distribution within cooling spheres, general case (also see Fig. 11.8a).

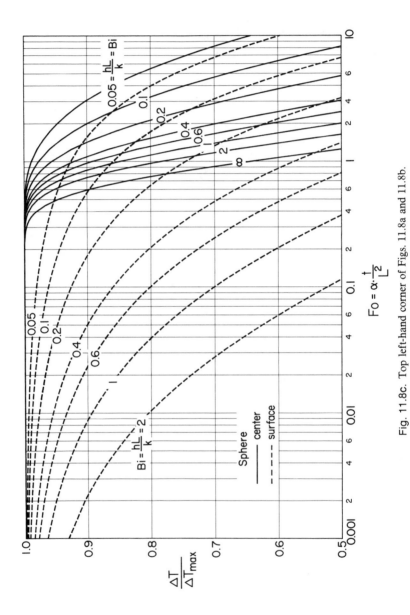

Fig. 11.8c. Top left-hand corner of Figs. 11.8a and 11.8b.

groups:

- an unaccomplished temperature change: $\dfrac{\Delta T}{\Delta T_{max}} = \dfrac{T_s - T_g}{T_{s0} - T_g}$

- the fraction of heat remaining in the solids: $\dfrac{Q}{Q_{max}}$

- a relative time: $Fo = \dfrac{\alpha t}{L^2}$

- a resistance ratio: $Bi = \dfrac{hL}{k_s}$

- a radius, or distance ratio: $\dfrac{r}{R}$ or $\dfrac{l}{L}$

Figures 11.6–11.8 are the Gurney–Lurie (1923) charts accurately redrawn by Colakyan *et al.* (1984) to represent the changing temperature distribution within the cooling solids. Unfortunately, there is no characteristic length which will allow these charts to collapse into one graph. Hence for irregular solids one must interpolate between these graphs.

Figure 11.9 represents the changing heat content of cooling particles whose sizes are measured by their characteristic lengths. Note that in a wide range of conditions the curves for spheres, cylinders and flat plates all collapse to a single curve. Thus, the cooling rate of any irregular shaped particle can be evaluated directly from this chart.

The curves on the left side of this graph approach the extreme where surface resistance is negligible (see Fig. 11.5), while the curves on the right side of the graph approach the extreme where surface resistance dominates (see Fig. 11.3). The sketch of Fig. 11.10 displays the relationship between these figures.

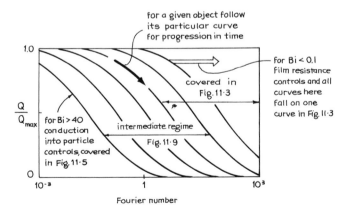

Fig. 11.10. Sketch of the relationship of the curves of Figs. 11.3, 11.5, and 11.9.

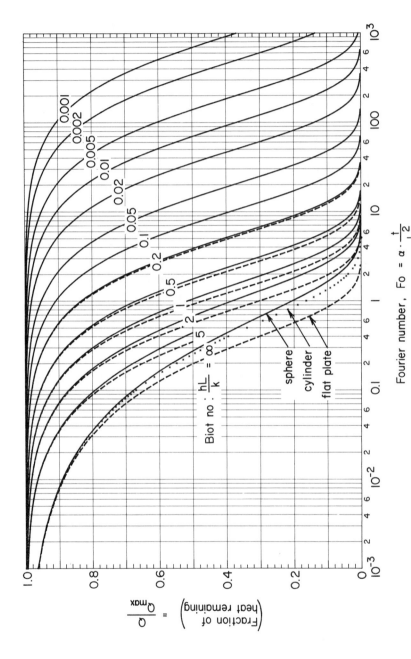

Fig. 11.9. General representation of the heat lost within a cooling sphere, infinite cylinder and infinite flat plate. [Prepared by Colakyan and Turton (1983).]

Fig. 11.11. The cooling of a semi-infinite object with negligible surface resistance.

IV. THE COOLING OF A SEMI-INFINITE SOLID FOR NEGLIGIBLE SURFACE RESISTANCE (Bi = $hL/k_s \rightarrow \infty$)

When a hot body at temperature T_{s0} is placed in contact with cold fluid at temperature T_g the surface immediately drops to T_g, heat flows from the body, and it progressively cools as shown in Fig. 11.11. The governing differential equation for conduction, equation (11.1), when integrated for the boundary conditions of this situation, gives the temperature at any time

Table 11.1. Values of the Error Function[a]
[This gives the solution to equation (11.12)]

$y = \dfrac{l}{\sqrt{4\pi t}}$	$\mathrm{erf}(y) = \dfrac{\Delta T}{\Delta T_{max}}$	$y = \dfrac{l}{\sqrt{4\pi t}}$	$\mathrm{erf}(y) = \dfrac{\Delta T}{\Delta T_{max}}$
0.0	0.0	0.70	0.678
0.05	0.056	0.75	0.711
0.10	0.112	0.80	0.742
0.15	0.168	0.85	0.771
0.20	0.223	0.90	0.797
0.25	0.276	0.95	0.821
0.30	0.329	1.0	0.843
0.35	0.379	1.2	0.910
0.40	0.428	1.4	0.952
0.45	0.476	1.6	0.976
0.50	0.520	1.8	0.989
0.55	0.563	2.0	0.995
0.60	0.604	2.5	0.9996
0.65	0.642	∞	1

[a]From *Tables of the Error Function and Its Derivative*, National Bureau of Standards, Applied Mathematics Series 41, Washington, D.C. (1954).

Fig. 11.12. Temperature distribution at any time and any position in a cooling semi-infinite object, general case [Equation (11.16)], and for no surface resistance [Equation (11.12)].

and at any point in the object in terms of the Gaussian error function, as follows [see Welty (1974)]:

$$\frac{\Delta T}{\Delta T_{\max}} = \frac{T_s - T_g}{T_{s0} - T_g} = \mathrm{erf}\left(\frac{l}{\sqrt{4\alpha t}}\right) = \mathrm{erf}(y) \qquad (11.12)$$

where the error function is defined as

$$\mathrm{erf}(y) = \frac{2}{\sqrt{\pi}} \int_0^y e^{-x^2} dx \qquad (11.13)$$

Numerical values for the error function are given in Table 11.1. From this the temperature of the solid can be evaluated directly at any position and at any time, as shown in the lowest curve of Fig. 11.12.

The instantaneous rate of heat loss is found from equation (11.12) by evaluating the temperature gradient at the surface (at $l = 0$). From the mathematics this gives

$$-\dot{q} = k_s A \frac{\partial T_s}{\partial l}\bigg|_{l=0} = k_s A\left(\frac{T_{s0} - T_g}{\sqrt{\pi\alpha t}}\right) \qquad [\mathrm{W}] \qquad (11.14)$$

Finally, the total amount of heat loss from time $t = 0$ is found by integrating the instantaneous rate of heat loss, equation (11.14), giving

$$Q_{\mathrm{lost}} = \int_0^t (-\dot{q})\, dt = 2 k_s A (T_{s0} - T_g) \sqrt{\frac{t}{\pi\alpha}} \qquad [\mathrm{J}] \qquad (11.15)$$

V. THE COOLING OF A SEMI-INFINITE BODY INCLUDING A SURFACE RESISTANCE

This situation is similar to the previous case treated, but with an added surface resistance. Thus we have a cooling behavior somewhat as sketched in Fig. 11.13. Integration of the conduction equation for this situation [see Sucec (1975)] gives the temperature of the solid at any position l from the surface at any time t as:

$$\frac{\Delta T}{\Delta T_{\max}} = \mathrm{erf}\left(\frac{1}{2\sqrt{\mathrm{Fo}}}\right) + \left[1 - \mathrm{erf}\left(\frac{1}{2\sqrt{\mathrm{Fo}}} + \mathrm{Bi}\sqrt{\mathrm{Fo}}\right)\right] \exp(\mathrm{Bi} + \mathrm{Bi}^2 \cdot \mathrm{Fo})$$

$$(11.16)$$

Fig. 11.13. The cooling of a semi-infinite object with surface resistance, the general case.

By putting Bi = ∞ we see that equation (11.16) reduces directly to the special case expression of equation (11.12).

The cooling curves for both finite and infinite values of the Biot number are shown in Fig. 11.12. Note that at any point in the object the approach to equilibrium (or the final temperature) takes a longer time for larger surface resistance (smaller values of the Biot number).

VI. HEAT LOSS IN OBJECTS OF SIZE L FOR SHORT COOLING TIMES

Short cooling times mean that heat is only lost in the outer layer of the solid, and that cooling has not yet penetrated deep into the object. In these situations the object can be treated as a semi-infinite solid, and its cooling behavior, whether it be slab, sphere, or what not, only depends on the amount of surface in contact with the fluid, not its shape.

Thus, equations (11.12) and (11.16) and Fig. 11.12 represent the cooling in this time period. Boelter et al. (1956) and Schack (1965) both find that these simple "short time" solutions apply in the time period

$$Fo = \frac{\alpha t}{L^2} < 0.077$$

where L is the characteristic length of the particle.

VII. THE COOLING OF FINITE OBJECTS SUCH AS CUBES, SHORT CYLINDERS, RECTANGULAR PARALLELOPIPEDS, AND SO ON

The temperature at any point in these finite objects is related to the corresponding temperature in the three mutually perpendicular infinite bodies whose intersections produce the object in question. This relationship is a simple one, and is given by Sucec (1975) as

$$\left(\frac{\Delta T}{\Delta T_{max}}\right)_{object} = \left(\frac{\Delta T}{\Delta T_{max}}\right)_x \left(\frac{\Delta T}{\Delta T_{max}}\right)_y \left(\frac{\Delta T}{\Delta T_{max}}\right)_z \qquad (11.17)$$

where the terms on the right are evaluated from Figs. 11.6 and/or 7 for the point in question.

Similarly, the total heat loss from a finite object is related to the heat loss from the bounding infinite slabs and cylinders by

$$\left(\frac{Q}{Q_{max}}\right)_{object} = \left(\frac{Q}{Q_{max}}\right)_x \left(\frac{Q}{Q_{max}}\right)_y \left(\frac{Q}{Q_{max}}\right)_z \qquad (11.18)$$

where the terms on the right are evaluated from Fig. 11.9.

Example 11.2 shows how to use these equations.

VIII. INTRUSION OF RADIATION EFFECTS

When heat enters or leaves a body by both convection and radiation the coefficient h used in the Biot numbers throughout this chapter should be the overall coefficient accounting for both these mechanisms of heat transfer, or

$$h_{overall} = h_{convection} + h_{radiation}$$

This radiation coefficient may change considerably as the surface temperature of the particle changes. To determine whether the radiation contribution is appreciable compared to convection and thus needs be considered, use Fig. 9.6 with a correction for view factor and emissivity.

IX. NOTE ON THE USE OF THE BIOT AND FOURIER NUMBERS

In most tests the Fourier and the Biot numbers for spheres and cylinders are defined in terms of the radius of the object, rather than the characteristic size of the object, V/A. Be careful not to confuse these measures in the charts from other books:

for flat plates: $\mathrm{Fo}_{here} = \mathrm{Fo}_{other}$; $\mathrm{Bi}_{here} = \mathrm{Bi}_{other}$

for cylinders: $\mathrm{Fo}_{here} = 4\mathrm{Fo}_{other}$; $\mathrm{Bi}_{here} = \frac{1}{2}\mathrm{Bi}_{other}$

for spheres: $\mathrm{Fo}_{here} = 9\mathrm{Fo}_{other}$; $\mathrm{Bi}_{here} = \frac{1}{3}\mathrm{Bi}_{other}$

The advantage of using V/A over R is that the curves for various shapes of objects often lie close to each other or collapse to a single curve. In addition, with the definition used here the Biot and Fourier numbers are the actual ratios of resistances as defined in equations (11.3) and (11.4).

REFERENCES AND NOTES

L. M. K. Boelter V. H. Cherry, H. A. Johnson, and R. C. Martinelli, *Heat Transfer Notes*, McGraw-Hill, New York (1956). Gives detailed derivations of the many equations for unsteady state conduction used in this chapter.

M. Colakyan, R. Turton, and O. Levenspiel, Unsteady-state transfer to various shaped objects, *Heat Transfer Engineering*, in press (1984).

H. Gröber, S. Erk, and U. Gringull, *Fundamentals of Heat Transfer* (translated from the German by J. R. Moszynski), McGraw-Hill, New York (1961). Also a good source book for many of the underlying equations of this chapter.

H. P. Gurney and J. Lurie, Charts for estimating temperature distributions in heating and cooling solid shapes, *Ind. Eng. Chem.* **15**, 1170 (1923).

J. Mator, M. S. Project, Chemical Engineering Department, Oregon State University, Corvallis (1982).

A. Schack, *Industrial Heat Transfer* (translated from the 6th German ed., by I. Gutman), Wiley, New York (1965).

J. Sucec, *Heat Transfer*, Simon and Schuster, New York (1975).

J. R. Welty, *Engineering Heat Transfer*, p. 135, Wiley, New York (1974).

EXAMPLE 11.1. VERIFYING A KEY ASSUMPTION IN THE ANALYSIS OF FLUIDIZED BED HEAT EXCHANGERS

When a stream of hot solids is contacted by cold gas in a fluidized bed heat exchanger the simple treatment of Chapter 14 assumes that:

(a) cold entering gas heats up instantaneously to the bed temperature,

(b) each particle of entering solid cools down instantaneously to the bed temperature, and

(c) both gas and solid leave the bed at the same temperature.

The term "instantaneous" as used above is reasonably approximated when the time needed for the two feed streams to get close to the bed temperature is much shorter than the time of stay of those streams in the exchanger. Preliminary considerations in Chapter 14 showed that assumption (a) is reasonably well met in practice. Let us look at assumption (b) here.

Suppose that hot sand (d_p = 600 μm) at 1000°C flows continuously into a fluidized bed exchanger where it is cooled by air at room temperature, and that gas and solid leave at about 500°C. For a fluidizing velocity of u_0 = 0.5 m/s at 1.1 atm find how long it takes for an incoming particle to cool to 550°C (90% approach).

(a) Assume that conduction within the particles controls.
(b) Assume that film resistance at the surface of the particles controls.
(c) Account for both resistances.

Compare these times to the mean residence time of solids in the exchanger, about three minutes.

Solution

The problem is to find how long it takes for a particle to cool such that Q/Q_{max} = 0.1. Let us tabulate all the physical properties needed to answer this question. For sand at 500°C Appendix U gives

$$\rho_s = 2600 \text{ kg/m}^3$$

$$k_s = 0.33 \text{ W/m K}$$

$$C_s = 800 \text{ J/kg K}$$

thus

$$\alpha = k_s/\rho_s C_s = 1.59 \times 10^{-7} \text{ m}^2/\text{s}$$

For air at 1.1 atm and 500°C, from Appendix U:

$$\mu_g = 36.19 \times 10^{-6} \text{ kg/m s}$$

$$k_g = 57.45 \times 10^{-3} \text{ W/m K}$$

$$C_g = 1093 \text{ J/kg K}$$

and from Appendix L

$$\rho_g = \frac{p_A(mw)}{RT} = \frac{(101{,}325)(1.1)(0.0289)}{(8.314)(773)} = 0.50 \text{ kg/m}^3$$

The heat transfer coefficient between a fluidized particle and its surroundings is given by equation (9.39). Replacing values gives

$$h = \frac{k_g}{d_p}\left[2 + 0.6\left(\frac{d_p u_0 \rho_g}{\mu_g}\right)^{1/2}\left(\frac{C_g \mu_g}{k_g}\right)^{1/3}\right]$$

$$= \frac{57.45 \times 10^{-3}}{6 \times 10^{-4}}\left\{2 + 0.6\left[\frac{(6 \times 10^{-4})(0.5)(0.5)}{36.19 \times 10^{-6}}\right]^{1/2}\left[\frac{(1093)(36.19 \times 10^{-6})}{57.45 \times 10^{-3}}\right]\right\}$$

$$= 295 \text{ W/m}^2 \text{ K}$$

We are now ready to solve for the cooling time based on the three different assumptions.

Assumption A. Particle Conduction Controls

For spherical particles $(L = d_p/6 = 10^{-4} \text{ m})$ and $Q/Q_{max} = 0.1$, Fig. 11.5 gives

$$\text{Fo} = \alpha \frac{t}{L^2} = 1.7$$

from which

$$t = 1.7 \frac{L^2}{\alpha}$$

Replacing values gives

$$t = \frac{(1.7)(10^{-4})^2}{1.59 \times 10^{-7}} = 0.11 \text{ s}$$

Assumption B. Film Resistance Controls

Method A. Use equation (11.8). From equation (11.8) rearranged we have

$$t = \frac{L\rho_s C_s}{h} \ln \frac{Q_{max}}{Q}$$

$$= \frac{(10^{-4})(2600)(800)}{295} \ln 10 = \boxed{1.62 \text{ s}}$$

Method B. Use Fig. 11.3. From the lowest curve of this figure we have

$$\text{Bi} \cdot \text{Fo} = \frac{ht}{L\rho_s C_s} = 2.32$$

Thus

$$t = 2.32\left(\frac{L\rho_s C_s}{h}\right)$$

$$= 2.32\left[\frac{(10^{-4})(2600)(800)}{295}\right] = \boxed{1.64 \text{ s}}$$

Assumption C. Accounting for Both Resistances

For this we must first evaluate the Biot number for the cooling particles. Thus

$$\text{Bi} = \frac{hL}{k_s} = \frac{(295)(10^{-4})}{0.33} = 0.0893$$

Then Fig. 11.9 shows that

$$\text{Fo} = \alpha \frac{t}{L^2} = 28$$

from which

$$t = 28 \frac{L^2}{\alpha} = \frac{28(10^{-4})^2}{1.59 \times 10^{-7}} = \boxed{1.76 \text{ s}}$$

Note that Fig. 11.3 can be used for this solution in place of Fig. 11.9. However, if you try using the design charts of Fig. 11.6 you will have problems.

Comments

As expected, the correct solution which accounts for both resistances gives a longer time, 1.76 s, than either of the solutions which only considers one or the other of the two resistances.

Comparing the two resistances, we see that the film dominates (1.64 s vs. 0.11 s), and just about controls (1.64 s vs. 1.76 s). Now both Figs. 11.1 and 11.10 point out that when Bi < 0.1, then one can assume that film resistance controls. In this problem Bi = 0.089, which just meets this condition, and so the results are as expected.

The cooling time found here (1.76 s) is very much shorter than the mean residence time in the exchanger (3 min), hence the assumption that the hot incoming particles cool instantaneously to the bed temperature is a reasonable idealization of the state of affairs.

Table 14.1 gives the relaxation times for a whole range of materials and particle sizes, calculated by the method of this example.

EXAMPLE 11.2. DEEP-FRIED FISH STICKS

A cod fillet, about $6 \times 1 \times 2$ cm, is taken from a cooler at 0°C and slipped in hot oil at 180°C.

(a) What is the centerpoint temperature of the fillet after 5 min?

(b) How much heat has been taken up by the fillet during this time?

Data: For cod:

$$k = 0.5 \text{ W/m K}$$

$$\alpha = 0.17 \times 10^{-6} \text{ m}^2/\text{s}$$

For the fillet in the deep-fat fryer:

$$h = 150 \text{ W/m}^2 \text{ K} \qquad \text{for the two small end faces}$$

$$h = 100 \text{ W/m}^2 \text{ K} \qquad \text{for the four long faces.}$$

Solution

The fillet can be represented by the intersection of three mutually perpendicular planes 6 cm, 1 cm, and 2 cm thick. Thus

$$L_x = 0.06/2 = 0.03 \text{ m}$$

$$L_y = 0.01/2 = 0.005 \text{ m}$$

$$L_z = 0.02/2 = 0.01 \text{ m}$$

The individual Biot numbers are

$$\text{Bi}_x = \frac{h_x L_x}{k} = \frac{150(0.03)}{0.5} = 9$$

$$\text{Bi}_y = \frac{100(0.005)}{0.5} = 1$$

$$\text{Bi}_z = \frac{100(0.01)}{0.5} = 2$$

The individual Fourier numbers are

$$\text{Fo}_x = \frac{\alpha t}{L_x^2} = \frac{(0.17 \times 10^{-6})(300)}{(0.03)^2} = 0.055$$

$$\text{Fo}_y = \frac{(0.17 \times 10^{-6})(300)}{(0.005)^2} = 2.0$$

$$\text{Fo}_z = \frac{(0.17 \times 10^{-6})(300)}{(0.01)^2} = 0.50$$

A. Centerpoint Temperature. From Fig. 11.6, for the three midplanes, we have

$$\left(\frac{\Delta T}{\Delta T_{\max}}\right)_x \cong 1$$

$$\left(\frac{\Delta T}{\Delta T_{\max}}\right)_y = 0.25$$

$$\left(\frac{\Delta T}{\Delta T_{\max}}\right)_z = 0.65$$

We are now ready to replace in equation (11.17). Thus, at the intersection of the three midplanes, in effect the centerpoint, we have

$$\left(\frac{\Delta T}{\Delta T_{\max}}\right)_{\text{centerpoint}} = \frac{T_{\text{oil}} - T_{\text{center}}}{T_{\text{oil}} - T_{\text{cooler}}} = \left(\frac{\Delta T}{\Delta T_{\max}}\right)_x \left(\frac{\Delta T}{\Delta T_{\max}}\right)_y \left(\frac{\Delta T}{\Delta T_{\max}}\right)_z$$

$$= 1(0.25)(0.65) = 0.1625 \tag{i}$$

but

$$\left(\frac{\Delta T}{\Delta T_{\max}}\right)_{\text{centerpoint}} = \frac{T_{\text{oil}} - T_{\text{center}}}{T_{\text{oil}} - T_{\text{cooler}}} = \frac{180 - T_{\text{center}}}{180 - 0} \tag{ii}$$

Combining equations (i) and (ii) gives

$$T_{\text{center}} = \boxed{151\,^\circ\text{C}}$$

B. Heat Absorbed. First of all, for cod fish fillets

$$\rho = 1050 \text{ kg/m}^3 \text{ (approximately)}$$

$$C_p = \frac{k}{\rho \alpha} = \frac{0.5}{(1050)(0.17 \times 10^{-6})} = 2801 \text{ J/kg K}$$

The maximum amount of heat which can be absorbed

$$Q_{\max} = W C_p (T_{\text{oil}} - T_{\text{cooler}})$$

$$= [(1050)(0.06 \times 0.01 \times 0.02)](2801)(180 - 0) = 6353 \text{ J} \tag{iii}$$

For the three intersecting infinite slabs, each with its own Biot and Fourier number, Fig. 11.9 shows that

$$\left(\frac{Q}{Q_{max}}\right)_x = 0.76 \quad \text{at Bi = 9 and Fo = 0.055}$$

$$\left(\frac{Q}{Q_{max}}\right)_y = 0.23 \quad \text{at Bi = 1 and Fo = 2}$$

$$\left(\frac{Q}{Q_{max}}\right)_z = 0.54 \quad \text{at Bi = 2 and Fo = 0.5}$$

Thus for the fish fillet, equation (11.18) becomes

$$\left(\frac{Q}{Q_{max}}\right) = (0.76)(0.23)(0.54) = 0.094 \qquad \text{(iv)}$$

and with equation (iii) the heat remaining to be absorbed is

$$Q = (0.094)(6353) = 997$$

So the heat which has been absorbed is

$$Q = 6353 - 997 = 5756 \text{ J}$$

NOTE: One can find the average temperature of the fillet directly from these Q values.

PROBLEMS ON UNSTEADY-STATE HEATING AND COOLING OF SOLID OBJECTS

11.1. Table 14.1 states that a 1-cm PVC plastic sphere moving through air at 1 m/s would have a thermal relaxation time of 170 s. Verify this figure.

11.2. A cold (0°C), long, instrumented copper cylinder 5-cm-o.d. is quickly plunged into a fluidized bed maintained at 100°C, and the cylinder's center point temperature reads 40, 60 and 80°C after 60, 110 and 200 s. Find the heat transfer coefficient between cylinder and bed.

11.3. *Roasting peanuts.* One method of preparing fat-free roasted peanuts involves lowering a wire basket of raw shelled peanuts into a vat of molten mannitol and sorbitol (nonsweet sugars) instead of into hot oil. When the peanuts are well roasted they are removed, drained, lightly salted and ready for packing.

If peanuts, originally at 15°C, are lowered into the 165°C roasting medium

(a) find the time needed for their centers to reach 105°C.

(b) how hot does the surface temperature of the peanuts become?

Data and Assumptions: Assume that the peanuts are close to spherical with diameters of 7.5 mm, and have the following properties

$$k_s = 0.5 \text{ W/m K}$$

$$\rho_s = 1150 \text{ kg/m}^3$$

$$C_s = 1700 \text{ J/kg K}$$

Between peanuts and molten sugar take $h = 80 \text{ W/m}^2 \text{ K}$.

11.4. *More on roasting peanuts.* Another way to dry-roast peanuts is to dump a batch of peanuts into a fluidized bed of mannitol particles kept at 140°C (melting point of mannitol = 160°C) and then remove them when their centers reach 105°C. This process does not leave a coating of hexose on the peanuts (see previous problem).

(a) Find the time needed to roast a batch of peanuts this way.

(b) Estimate the surface temperature of the just roasted peanuts.

Data: See previous problem for the thermal properties of peanuts. Between peanuts and fluidized bed take

$$h = 200 \text{ W/m}^2 \text{ K}$$

11.5. The thermal properties of a sand pile (voidage = 0.4) is to be evaluated by quickly pouring a bucketful of hot sand into a 10-cm-i.d. water-cooled aluminum pipe having a centrally placed thermocouple. The thermocouple reads 140°C just after the sand is poured in, 32°C after 25 min, 25°C after 33 min, and 20°C after a long time.

From this information evaluate the mean thermal conductivity and mean thermal diffusivity of a sand pile.

11.6. *Restarting a fluidized bed incinerator.* If the air flow to a fluidized bed incinerator stops because of a power outage or some such reason, the solids will collapse to form a "slumped" bed, which then will cool slowly. If the bed is still hot enough when the flow of cold fluidizing air is resumed it will reignite spontaneously. However, if the bed temperature drops below the ignition temperature then a long involved procedure must be followed to reignite it.

Let us estimate how long the bed can remain slumped, and still restart spontaneously when air flow is resumed. In restarting:

(a) assume that only a little air is fed to the bed at the beginning, so that the bed remains slumped while the hot spot reignites and spreads. Then air flow is turned up and the bed refluidizes.

(b) assume that full air flow is used immediately so that the slumped bed refluidizes right away.

Can you think of any of the advantages and drawbacks of these alternatives?

Data: The slumped bed is 1 m deep and 4 m in diameter. The hot operating fluidized bed is at 850°C. The ignition temperature is 600°C. The thermal diffusivity of the slumped bed is 10^{-6} m^2/s. Assume that the top and bottom surfaces of the slumped bed drop to room temperature in about an hour but that the bed walls are well insulated.

11.7. *Heating of carbon particles*. Hot spherical carbon particles ($d_p = 3$ mm) are needed for an experiment. To prepare this feed material, carbon spheres at 0°C are allowed to fall one at a time through a large diameter heated pipe containing nitrogen at 1 atm. Wall and gas are kept at 500°C.

(a) Determine the length of pipe needed for the leaving particles to be at an average temperature of 300°C.

(b) What is the surface temperature and the midpoint temperature of these leaving particles?

Data: For carbon particles:

$$\rho_s = 550 \text{ kg/m}^3$$

$$C_s = 1415 \text{ J/kg K}$$

$$k_s = 0.18 \text{ W/m K}$$

Assume that the particles are falling at their terminal velocity when they enter the heating chamber.

11.8. A very viscous chocolate syrup is to be heated by forcing it through a scraped wall heat exchanger. If the blades wipe the walls clean and bring fresh syrup to the surface twice a second estimate the heat transfer coefficient at the walls of the exchanger.

Data: Thermal properties of chocolate syrup are estimated to be

$$k_s = 0.5 \text{ W/m K}$$

$$\rho_s = 1200 \text{ kg/m}^3$$

$$C_s = 3600 \text{ J/kg K}$$

Chapter 12 INTRODUCTION TO
HEAT EXCHANGERS

Heat exchangers are devices for transferring heat from a hot flowing stream to a cold flowing stream. There are three broad types of exchangers:

- the recuperator, or the through-the-wall nonstoring exchanger;
- the direct-contact nonstoring exchanger;
- the regenerator, accumulator, or heat storage exchanger.

The type chosen in any situation depends in large part on the nature of the transferring phases, whether gas–gas, gas–liquid, gas–solid, liquid–liquid, liquid–solid, or solid–solid, and on the mutual solubility of the phases involved.

Let us present some examples of these three kinds of exchangers.

I. RECUPERATORS (THROUGH-THE-WALL NONSTORING EXCHANGERS)

In recuperators the two flowing streams are separated by a wall and heat has to pass through this wall. Many different contacting patterns are

Fig. 12.1. Various kinds of recuperators, or through-the-wall exchangers.

used for the two fluids and we will study a number of them in later chapters. The sketches of Fig. 12.1 indicate some of these many different contacting patterns.

Recuperators are certainly less effective than the direct contact exchanger because the presence of the wall hinders the flow of heat. But this type of exchanger is used where the fluids are not allowed to contact each other, as with gas–gas systems, miscible liquids, dissolving solids, or reactive chemicals.

II. DIRECT-CONTACT NONSTORING EXCHANGERS

In direct-contact nonstoring exchangers the streams contact each other intimately, the hotter stream giving up its heat directly to the colder stream.

MOVING BED
CONTACTOR

FLUIDIZED
BED

MOVING BELT
CONVEYOR

Fig. 12.2. Gas–solid direct contact nonstoring exchangers.

Naturally this type of exchanger finds use when the two contacting phases are mutually insoluble and do not react with each other. Thus, it cannot be used with gas–gas systems.

Direct-contact exchangers are of three broad types. First there are the gas–solid exchangers, various forms of which are shown in Fig. 12.2.

Then there are the fluid–fluid exchangers where the two contacting fluids are mutually immiscible. These are shown in Fig. 12.3.

Finally, it is not always necessary that the two contacting fluids be mutually insoluble, and Fig. 12.4 shows exchangers where one of the flowing fluids does dissolve in the other. In particular, in air–water systems the direct contact exchanger is of prime importance just because one of the phases (water) does dissolve, or evaporate, in the other phase (air). The

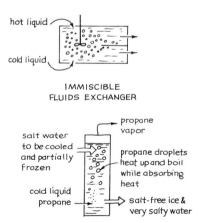

IMMISCIBLE
FLUIDS EXCHANGER

Fig. 12.3. Fluid–fluid direct-contact nonstoring exchangers.

BOILING & IMMISCIBLE
FLUIDS EXCHANGER

Fig. 12.4. Fluid–fluid direct-contact exchangers where one phase can dissolve in the other.

water cooling tower, shown in Fig. 12.4, is such an example, and in fact, represents the most widely used type of heat exchanger in industry.

The proper treatment of this type of exchanger requires using the methods of simultaneous heat and mass transfer, and are beyond the scope of this volume. The interested reader is referred to Fair (1972a; 1972b) and to the many standard books on unit operations and mass transfer for such a presentation.

III. REGENERATORS (DIRECT-CONTACT HEAT STORING EXCHANGERS)

With regenerators a hot stream of gas transfers its heat to an intermediary, usually a solid, which later gives up this stored heat to a second stream of cold gas. There are a number of different ways of doing this, as shown in Fig. 12.5.

Fig. 12.5. Heat regenerators or heat storing exchangers: (a) Heat storing solids are stationary; (b) heat storing solids continuously circulate between hot and cold streams.

Fig. 12.6. The heat pipe transfers heat from one place to another, often far apart.

IV. EXCHANGERS USING A GO-BETWEEN STREAM

In a number of difficult situations or when the two heat exchanging locations are far apart, a third stream may be used as go-between, to take up heat from the hot stream and then deliver it to the cold stream. This go-between stream may consist of solid particles or it may be a fluid. Consider a few processes which use this idea.

A. The Heat Pipe for Heat Exchange at a Distance

The heat pipe (see discussion after Problem 3.24) very effectively transports heat from one location to another, and since the primary resistance to heat transfer is at the two ends of the pipe, where heat is taken up

Fig. 12.7. The SPHER process [J. E. Gwyn *et al.*, *Chem. Eng. News*, p. 42 (Sept. 15, 1980)] for the recovery of heat from spent shale, and its transfer to cold fresh shale. This is a counterflow solid–solid heat exchanger which uses a third stream of solids as go-between.

and released, finned tubes are normally used in these sections, as shown in Fig. 12.6. The fluid in the pipe which boils at one end and condenses at the other acts as the go-between in the transfer of heat.

Examples of the use of heat pipes are in the recovery of stored heat in solar home heating (see problem 3.27), in space capsules as a means of transferring heat from the hot side (facing the sun) to the cold side of the capsule, and in microelectronic and hi-fi equipment to draw off heat from critical components and dissipate it into the air, thereby avoiding overheating. Finally, close to 100,000 heat pipes were built into the Alaska pipeline supports to keep heat away from the support footings which had to be imbedded in permafrost soil.

B. Solid–Solid Heat Transfer

To achieve counterflow heat exchange of gases and liquids is no problem, but for two streams of solids this is not an easy matter. Figure 12.7 shows one proposal which uses a recirculating carrier solid which falls countercurrent to the two primary solids in their separate fluidized beds. This is a complex operation, and the vigorous backmixing of solids in the fluidized beds results in severe deviations from the desired ideal of countercurrent plug flow of solids.

Another much simpler way for approaching countercurrent heat exchange of two streams of solids is to use a liquid stream as go-between, as shown in Fig. 12.8.

Still another approach uses the highly efficient heat pipes in a crisscross arrangement to yield countercurrent heat transfer. This is shown in Fig. 12.9.

Fig. 12.8. Counterflow solid–solid heat exchanger using a liquid go-between.

Fig. 12.9. Counterflow solid–solid heat exchanger using properly arranged heat pipes as go-between. [O. Levenspiel and R. T. Chan, U.S. Pat. No. 4,408,656.]

V. COMMENTS

What these designs show is that there are many different ways of transferring heat from one flowing stream to another, and the important first decision is to choose the right type of exchanger. Often this is a clean-cut decision, but sometimes one has to compare the economics of quite different kinds of exchangers.

In the following chapters we consider in turn the various kinds of exchangers, their important design parameters, and simple design methods.

REFERENCES

J. R. Fair, Process heat transfer by direct fluid phase contact, *AIChE Symp. Ser.* **68**, (118), 1 (1972a).

J. R. Fair, Designing direct-contact cooler/condensers, *Chem. Eng.*, p. 91 (June 12, 1972b).

Chapter 13   RECUPERATORS: THROUGH-THE-WALL NONSTORING EXCHANGERS

Recuperators are heat exchangers wherein the two fluids transferring heat are kept away from each other by a wall. Their design requires two pieces of information.

1. *The overall heat transfer coefficient U* which accounts for the overall resistance to transfer caused by the wall. This includes the individual film resistances and the wall resistance. From Chapter 10 we have

$$\frac{1}{UA} = \frac{1}{h_1 A_1} + \frac{\Delta x}{kA_{\text{mean}}} + \frac{1}{h_2 A_2}$$

2. *The contacting pattern of the two phases.* This can be very complex and difficult to evaluate. To make the analysis of exchangers tractable we

239

idealize real contacting patterns by considering them to be combinations of: (i) plug flow, (ii) mixed flow, (iii) batch completely mixed fluid, and (iv) batch unmixed and stagnant fluid.

Let us use the following nomenclature for this chapter:

A	area of exchanger [m^2]
C	specific heat of a fluid [J/kg K]
$\mathscr{F}$	$\overline{\Delta T}/\Delta T_{lm}$, mean temperature driving force in the exchanger compared to that of countercurrent flow, which is the very best possible [−]
$\dot{m}$	flow rate of a fluid [kg/s]
NTU	number of transfer units, based on a particular phase [−]; see Fig. 13.14
P	$\Delta T_i/\Delta T_{max}$, temperature change of phase i compared to the maximum possible temperature change [−]; see Fig. 13.4
$\dot{q}$	rate of heat gain by a phase [W]
Q	heat loss or gained by a fluid up to a given point in the exchanger [J/kg of one of the phases]
R	ratio of temperature changes or heat flows of the two fluids [−]; see Fig. 13.4
$\overline{\Delta T}$	proper mean driving force (or temperature difference) between the fluids [K]
ΔT_g	temperature change of gas phase [K]
ΔT_l	temperature change of liquid phase [K]
ΔT_{lm}	logarithmic mean temperature driving force [K]; see equation (13.5)
ΔT_{max}	maximum temperature difference between phases [K]

Subscripts:

0	at zero time
in	at entering conditions
out	at leaving conditions
g, l	we call the two contacting phases gas and liquid for convenience, though they may not be that
tx	transferred

For shell and tube exchangers, see equation (13.17):

- $\dot{M}$, C, and T refer to the shell-side flow rate, specific heat, and temperature, respectively;
- $\dot{m}$, c, and t refer to the tube-side flow rate, specific heat, and temperature, respectively.

Let us now develop performance equations for various contacting patterns.

I. COUNTERCURRENT AND COCURRENT PLUG FLOW

These two contacting patterns are idealization and approximations of the flows in double pipe heat exchangers. To develop the performance equation for these contacting patterns, and in general for any other contacting pattern, we draw on the following relationship:

$$
\begin{pmatrix} \text{Rate of} \\ \text{heat loss} \\ \text{by gas} \end{pmatrix} = \begin{pmatrix} \text{rate of} \\ \text{heat gain} \\ \text{by liquid} \end{pmatrix} = \begin{pmatrix} \text{rate of heat} \\ \text{transfer from} \\ \text{gas to liquid} \end{pmatrix} \qquad (13.1a)
$$

or

$$
\dot{q}_{\text{leaving gas}} = \dot{q}_{\text{entering liquid}} = \dot{q}_{\text{transferred}} \qquad (13.1b)
$$

We treat two broad cases; first, fluids with no phase change and with constant specific heat and, second, exchangers wherein phase changes such as boiling or condensation do occur.

A. No Phase Change, C_p Independent of Temperature

Figure 13.1 sketches the distance versus temperature diagram and the more useful Q vs. T diagram for these two idealized flow patterns. Note that Q represents the amount of heat exchanged as the fluids pass through

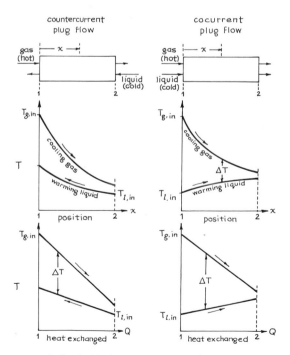

Fig. 13.1. Temperature profiles in ideal countercurrent and cocurrent exchangers, as a function of position and as a function of the heat transferred between streams.

the exchanger. This quantity is conveniently measured either by the enthalpy change of one of the fluids [J/kg], or as a fraction of the total heat transferred.

1. Countercurrent exchangers. Around the whole exchanger the heat balance of equation (13.1) can be written as

$$\dot{q} = -\dot{m}_g C_g \left(T_{g,\text{out}} - T_{g,\text{in}} \right) = \dot{m}_l C_l \left(T_{l,\text{out}} - T_{l,\text{in}} \right) = UA\overline{\Delta T} \quad (13.2)$$

Similarly, for a differential slice of exchanger of interfacial area dA, we have

$$d\dot{q} = \underbrace{-\dot{m}_g C_g \, dT_g}_{\text{I}} = \underbrace{\dot{m}_l C_l \, dT_l}_{\text{II}} = \underbrace{U\Delta T \, dA}_{\text{III}} \quad (13.3)$$

where $\Delta T = T_g - T_l$ is called the temperature driving force. Combining terms I and III, II and III, adding the results together and integrating while keeping U constant gives, after considerable manipulation,

$$\dot{q} = \dot{m}_g C_g \left(T_{g,\text{in}} - T_{g,\text{out}} \right) = \dot{m}_l C_l \left(T_{l,\text{out}} - T_{l,\text{in}} \right) = UA\Delta T_{lm} \quad (13.4)$$

where the proper temperature driving force is the logarithmic mean temperature driving force, defined as

$$\Delta T_{lm} = \ln \frac{\Delta T_2 - \Delta T_1}{\dfrac{\Delta T_2}{\Delta T_1}} = \ln \frac{\left(T_{g,\text{out}} - T_{l,\text{in}} \right) - \left(T_{g,\text{in}} - T_{l,\text{out}} \right)}{\dfrac{\Delta T_{g,\text{out}} - T_{l,\text{in}}}{T_{g,\text{in}} - T_{l,\text{out}}}} \quad (13.5)$$

Rearranging equations (13.4) and (13.5) gives, after even more manipulation, the efficiency of heat exchange as

$$\eta_l = \frac{\Delta T_l}{\Delta T_{\text{max}}} = \frac{T_{l,\text{out}} - T_{l,\text{in}}}{T_{g,\text{in}} - T_{l,\text{in}}} = \frac{1 - K}{\left(\dot{m}_l C_l / \dot{m}_g C_g \right) - K} \quad (13.6)$$

or

$$\eta_g = \frac{\Delta T_g}{\Delta T_{\text{max}}} = \frac{T_{g,\text{in}} - T_{g,\text{out}}}{T_{g,\text{in}} - T_{l,\text{in}}} = \frac{1 - K}{1 - \left(\dot{m}_g C_g / \dot{m}_l C_l \right) K} \quad (13.7)$$

where

$$K = \exp \left[-UA \left(\frac{1}{\dot{m}_g C_g} - \frac{1}{\dot{m}_l C_l} \right) \right] \quad (13.8)$$

In the special case of equal heat flows ($\dot{m}_g C_g = \dot{m}_l C_l = \dot{m}C$) the operating lines on the Q vs. T chart become parallel and

$$\Delta T_{lm} \rightarrow \Delta T \qquad \text{a constant throughout the exchanger}$$

In this situation equations (13.6) and (13.7) reduce to

$$\eta_g = \eta_l = \frac{\Delta T_g}{\Delta T_{max}} = \frac{\Delta T_l}{\Delta T_{max}} = \frac{UA/\dot{m}C}{1 + (UA/\dot{m}C)} \qquad (13.9)$$

Equations (13.4) and (13.5) are useful for finding the size of an exchanger given the desired inlet and outlet temperatures. Equations (13.6) to (13.9) are useful for the inverse problem, to find the outlet temperatures in a given system.

2. Cocurrent exchangers. Similar to countercurrent contacting, the size of exchanger (or flow rate) to give desired terminal temperatures is found to be

$$\dot{q} = \dot{m}_g C_g (T_{g,\,in} - T_{g,\,out}) = \dot{m}_l C_l (T_{l,\,out} - T_{l,\,in}) = UA\Delta T_{lm} \qquad (13.10)$$

where

$$\Delta T_{lm} = \ln \frac{\Delta T_2 - \Delta T_1}{\dfrac{\Delta T_2}{\Delta T_1}} = \ln \frac{(T_{g,\,out} - T_{l,\,out}) - (T_{g,\,in} - T_{l,\,in})}{\dfrac{T_{g,\,out} - T_{l,\,out}}{T_{g,\,in} - T_{l,\,in}}} \qquad (13.11)$$

From these expressions the efficiency of heat exchange (and thus the terminal temperatures) in a given exchanger is found to be

$$\eta_l = \frac{T_{l,\,out} - T_{l,\,in}}{T_{g,\,in} - T_{l,\,in}} = \frac{\dot{m}_g C_g}{\dot{m}_g C_g + \dot{m}_l C_l}(1 - K') \qquad (13.12)$$

or

$$\eta_g = \frac{T_{g,\,in} - T_{g,\,out}}{T_{g,\,in} - T_{l,\,in}} = \frac{\dot{m}_l C_l}{\dot{m}_g C_g + \dot{m}_l C_l}(1 - K') \qquad (13.13)$$

where

$$K' = \exp\left[-UA\left(\frac{1}{\dot{m}_g C_g} + \frac{1}{\dot{m}_l C_l} \right) \right] \qquad (13.14)$$

For equal heat flows ($\dot{m}_g C_g = \dot{m}_l C_l = \dot{m}C$) the above efficiency expressions reduce to

$$\eta_g = \eta_l = \frac{\Delta T_g}{\Delta T_{\max}} = \frac{\Delta T_l}{\Delta T_{\max}} = \frac{1}{2}[1 - e^{-(UA/2\dot{m}C)}] \qquad (13.15)$$

3. Cocurrent or countercurrent contacting with changing *U*. If *U* varies linearly with temperature integration of equation (13.3) gives

$$(U\Delta T)_{lm} = \ln\frac{U_2\Delta T_1 - U_1\Delta T_2}{\dfrac{U_2\Delta T_1}{U_1\Delta T_2}} \qquad \text{with } U = a + bT \qquad (13.16)$$

to be used in equation (13.4) or equation (13.10).

4. Conclusion. Countercurrent flow is always more efficient than cocurrent flow.

B. Exchangers with a Phase Change

Suppose cold water is used to condense hot atmospheric steam. For countercurrent flow the *Q* vs. *T* diagram will then be as shown in Fig. 13.2 with three distinct regimes, each with its particular *U* value. For this situation the overall heat loss/heat gain expression, equation (13.2), remains unchanged; however, the rate expression, equation (13.3), must be integrated either graphically, numerically, or by considering the three sections

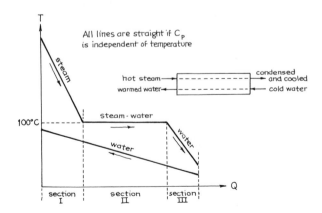

Fig. 13.2. *Q* vs. *T* diagram for countercurrent plug flow with a phase change.

of the exchanger separately, while using equations (13.4) and (13.5) for each section.

The Q vs. T graph is a useful representation for phase change systems, for finding intermediate conditions, and for helping one see what is happening. The values of Q for this graph can be found directly from the enthalpy tables for the fluids because in the absence of shaft work the energy balance for a flowing fluid, equation (1.4), reduces to

$$Q = \Delta H$$

Of course, the lines representing the hot fluid and the cold fluid can never cross each other on the Q vs. T graph.

II. SHELL AND TUBE EXCHANGERS

With shell and tube exchangers we have various flow patterns besides simple cocurrent and countercurrent plug flow. All are less efficient than countercurrent plug flow. By this we mean that they need more exchanger surface area for given end conditions. Despite this drawback these exchangers are widely used in industry because they often are more convenient, more compact, and less expensive to construct for a given duty.

We treat these exchangers as countercurrent plug flow units with a fudge factor $\mathscr{F}$, somewhere between 0 and 1, to account for the lowered contacting efficiency. Thus the performance equations for these exchangers, assuming constant U, becomes

$$\dot{q} = -\underbrace{\left[\dot{m}C(T_{out} - T_{in})\right]_{shell}}_{\text{Lost by shell fluid}} = \underbrace{\left[\dot{m}C(T_{out} - T_{in})\right]_{tube}}_{\text{Gained by tube fluid}} = \underbrace{UA\Delta T_{lm}\mathscr{F}}_{\substack{\text{Between shell} \\ \text{and tube-side fluids}}}$$

$$(13.17a)$$

or in the special nomenclature of this section

$$\dot{q} = \dot{M}C(T_2 - T_1) = \dot{m}c(t_2 - t_1) = UA\overline{\Delta T} \qquad (13.17b)*$$

There are all sorts of contacting patterns, designated by pairs of numbers 1-2, 1-4, 2-4, and so on. The first number refers to the number of shell

---

*Note that upper case letters $\dot{M}$, C, and T represent the shell-side fluid, lower case letters $\dot{m}$, c, and t represent the tube-side fluid, $\Delta T = T - t$ represents the temperature driving force, and subscripts 1 and 2 stand for the entering and leaving fluid stream, respectively.

Fig. 13.3. The insides of a 1-2 shell and tube exchanger, considerably simplified.

passes, the second refers to the number of tube passes, where the term "pass" represents flow from one end of the exchanger to the other.

A sketch of a 1-2 shell and tube exchanger is shown in Fig. 13.3. Such exchangers are often very large, containing not just six tubes, as shown here, but over a thousand tubes.

Equation (13.17) has been solved for a variety of shell and tube arrangements based on the following assumptions:

- ideal plug flow of all streams;
- no temperature gradients across the flow path of any fluid;
- same heat transfer area for each pass;
- constant U value throughout the exchanger;
- constant specific heat and no phase change for the fluids, hence no vaporization or condensation within the exchanger;
- negligible heat losses to the surroundings; and
- no conduction along the exchanger walls in the directions of flow of fluids.

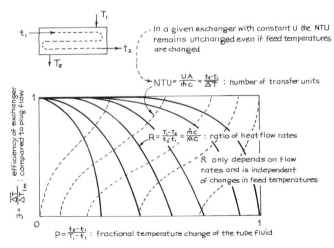

Fig. 13.4. Interrelationship of variables of equation (13.17) for shell and tube exchangers. $\dot{M}$, C, and T refer to the shell-side fluid; $\dot{m}$, c, and t refer to the tube-side fluid; $\Delta T = T - t$ is the temperature driving force.

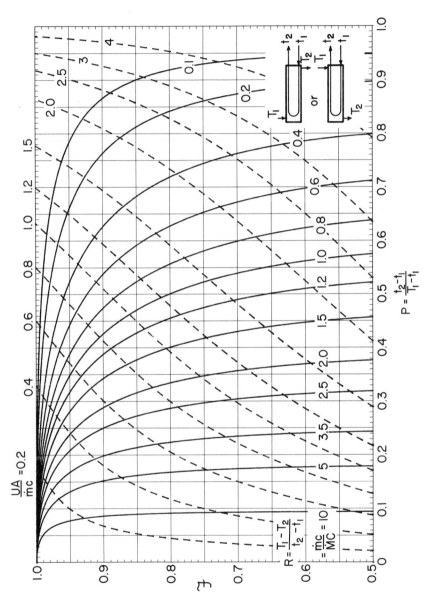

Fig. 13.5. Performance chart for a 1-2 shell and tube exchanger.

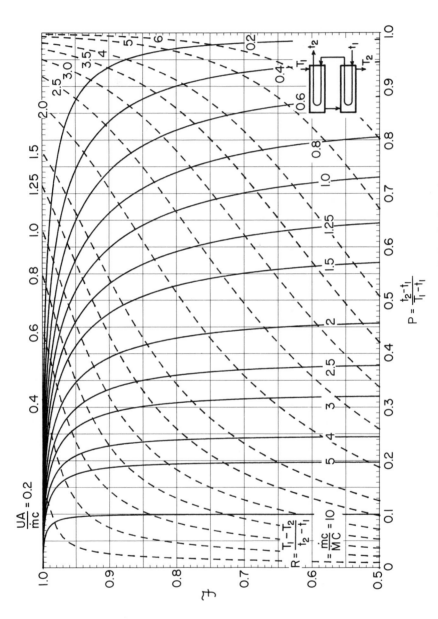

Fig. 13.6. Performance chart for a 2-4 shell and tube exchanger.

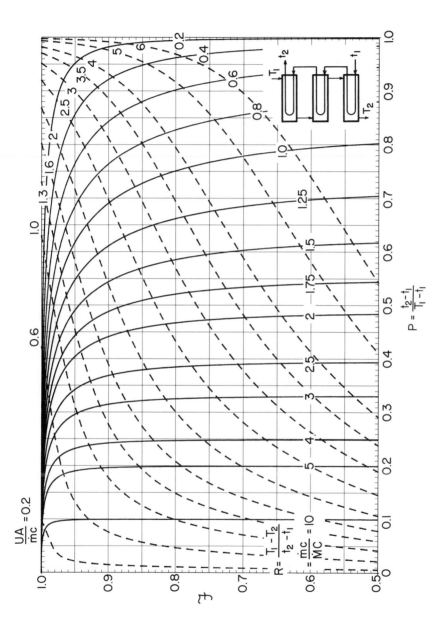

Fig. 13.7. Performance chart for a 3-6 shell and tube exchanger.

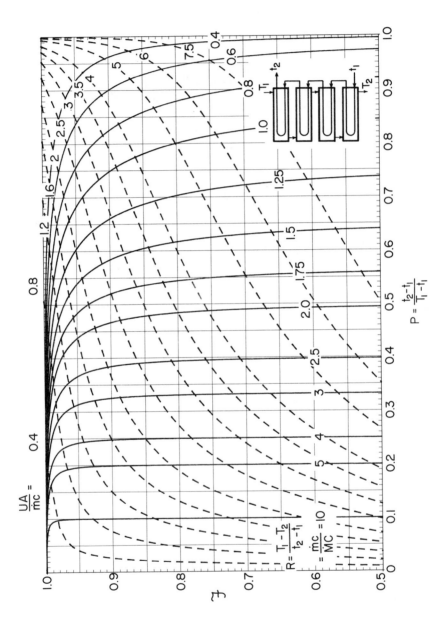

Fig. 13.8. Performance chart for a 4-8 shell and tube exchanger.

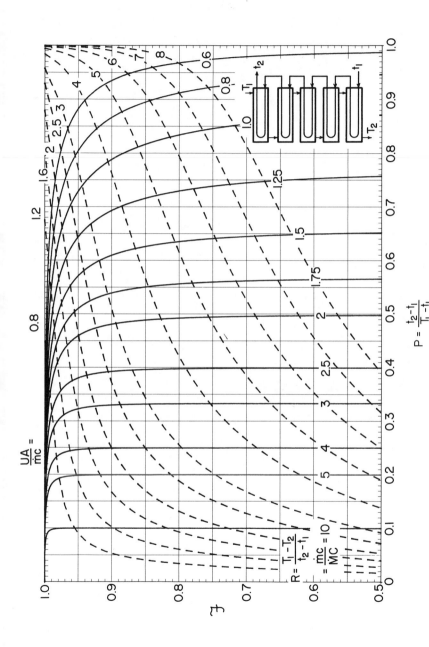

Fig. 13.9. Performance chart for a 5-10 shell and tube exchanger.

Fig. 13.10. Performance chart for a 6-12 shell and tube exchanger.

The resulting expressions are cumbersome but with the aid of computer graphics the relationship between all the variables can be displayed compactly in a chart of the type shown in Fig. 13.4. Figures 13.5 to 13.10, prepared by Turton *et al.* (1984), show this relationship for various combinations of shell and tube passes.

Comments

1. Shell and tube exchangers are used primarily for liquid–liquid systems.

2. Routing of fluids: Here are some considerations:

- The inside of tubes are much easier to clean than the shell side so the scum- or scale-forming fluid should flow through the tubes.
- The corrosive fluid should flow through the tubes to avoid the expense of special metals for both shell and tubes.
- The tube-side Δp usually is greater than the shell-side Δp, so the less viscous fluid should pass through the tubes.
- An odd number of tube passes, 3, 5, or 7, is rarely used because the expansion and stress problems associated with temperature changes are hard to deal with.
- There are various ways of contacting the phases for a given number of shell passes and tube passes. Always try to approach countercurrent flow.
- These charts show that the larger the number of shell and tube passes the closer does flow approach ordinary countercurrent flow with its largest temperature driving force.

3. The actual design of shell and tube exchangers is a complex business. The interested reader should look up the specialized books devoted to this subject.

III. CROSSFLOW AND COMPACT EXCHANGERS

These exchangers come in all shapes and sizes as sketched in Fig. 13.11. Here we recognize two sorts of flow for a phase: either well mixed laterally, or unmixed. For example, consider the temperature of the hot liquid at position A and at position B in the sketch of Fig. 13.12. With no lateral mixing, hot fluid will take separate and parallel paths through the exchanger

Fig. 13.11. Some of the many kinds of cross flow and compact heat exchangers.

in which case the fluid at B will be much cooler than the fluid at A. With lateral mixing all the fluid along AB will be at one and the same temperature. We show these two schemes as in Fig. 13.13.

The interrelationship of variables for various combinations of ideal mixed and ideal unmixed flow has been evaluated theoretically and has been displayed in design charts similar to Fig. 13.4. Figures 13.14–13.18, prepared by Turton *et al.* (1984), present accurate computer-drawn charts for various combinations of mixed and unmixed crossflow exchangers. In these charts, M, C, and T represent one fluid, while m, c, and t represent the other fluid. Comparing $\mathscr{F}$ values from Fig. 13.14, 13.15, and 13.16, shows that unmixed flow is better than mixed flow; and comparing Fig. 13.15 with 13.17 and 13.18 shows that multipass contacting is superior to single-pass contacting, if done properly (Fig. 13.17 vs. Fig. 13.15), but is poorer than single-pass contacting, if not done properly (Fig. 13.18 vs. Fig. 13.15).

There are many extensions to these contacting patterns. Mueller (1973) presents some of these. Figure 13.19 shows a few others.

Remarks

The sketches of Fig. 13.20 show four other ways of displaying the interrelationship of variables in heat exchangers. Sketch (a) is identical to the design charts of this chapter except that the variable NTU is left out.

Fig. 13.12   Fig. 13.13

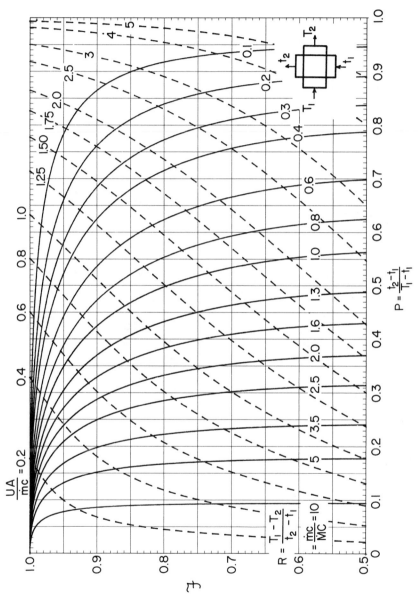

Fig. 13.14. Performance chart for a single-pass crossflow exchanger, both fluids mixed.

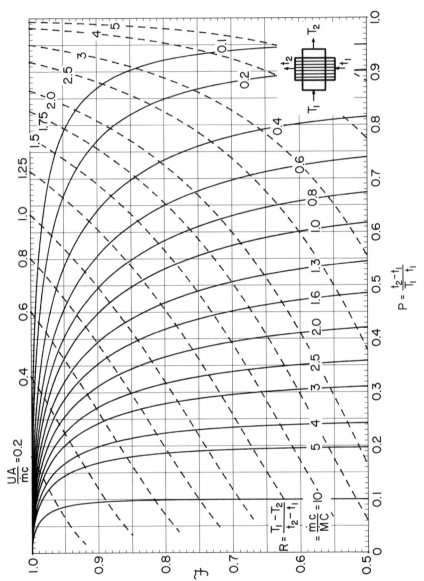

Fig. 13.15. Performance chart for a single-pass crossflow exchanger, one fluid mixed, one unmixed.

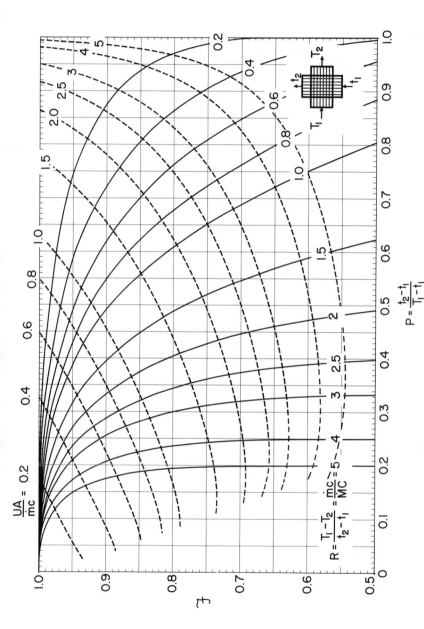

Fig. 13.16. Performance chart for a single-pass crossflow exchanger, both fluids unmixed. [See Baclic (1978) for the mathematics of this contracting pattern.]

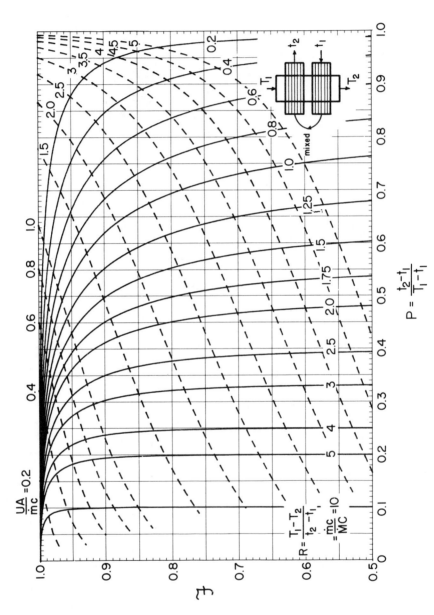

Fig. 13.17. Performance chart for a two-pass crossflow exchanger, countercurrent arrangement.

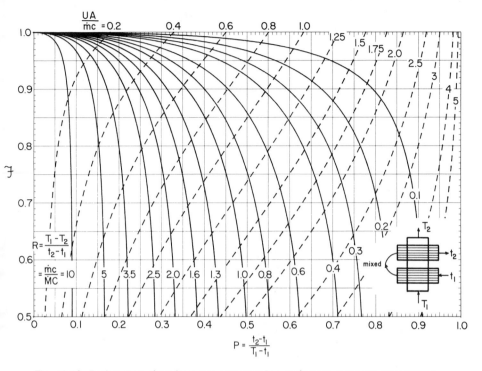

Fig. 13.18. Performance chart for a two-pass crossflow exchanger, cocurrent arrangement.

Consequently, although it allows one to find the size of exchanger directly, given the terminal temperatures of the exchanger, the reverse problem of finding the exit temperature of fluids in a given exchanger requires a tedious trial-and-error solution. Sketches (b) and (c) are essentially equivalent to charts of this chapter, in that they allow one to solve any kind of problem without trial and error. However, these figures do not display the $\mathscr{F}$ values,

Fig. 13.19

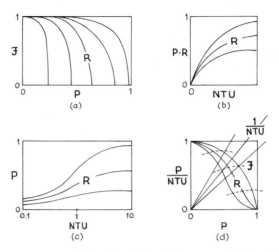

Fig. 13.20. Various ways of displaying the relationship between variables of a heat exchanger: (a) from Bowman *et al.* (1940), also presented in TEMA (1978); (b) from Kays and London (1963); (c) shown in TEMA (1978); (d) from Muller (1973).

which must be found with an extra calculation. Sketch (d) gives the same information as the charts here, however, its graph is not as spread out, and thus is harder to read.

IV. COLD FINGERS OR BAYONET EXCHANGERS

A cold finger or bayonet type of heat exchanger is used in blast furnaces and as a condenser in vacuum systems where proper sealing of the

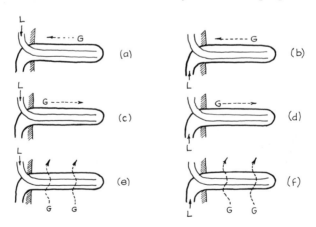

Fig. 13.21. Various possible contacting patterns for bayonet exchangers.

vessel to the inlet and outlet tubes may cause difficulties. There are six ideal contacting patterns for bayonet exchangers as shown in Fig. 13.21.

Let us evaluate the proper mean driving force $\overline{\Delta T}$ knowing the temperatures of the incoming streams. This value is then introduced into the performance expression:

$$\dot{q} = U_i A_i \overline{\Delta T} = U_o A_o \overline{\Delta T} \qquad (13.18)$$

We adopt the following nomenclature:

g	hot gas flowing on the outside
l	cool liquid flowing in the bayonet
T_l'	temperature in the inner tube
T_l''	temperature in the annulus
A_i, A_o	surface area of inner and of outer tube
P_i, P_o	perimeter of inner and of outer tube

Now consider arrangement (a), as shown in Fig. 13.22. Ignoring heat losses at the ends of the exchanger we have:

Rate of heat transfer into annulus

$$\left(T_g - T_l''\right)U_o P_o\, dx + \left(T_l' - T_l''\right)U_i P_i\, dx = -\dot{m}_l C_l\, dT_l'' \qquad (13.19)$$

Rate of heat transfer into inner tube

$$\left(T_l'' - T_l'\right)U_i P_i\, dx = \dot{m}_l C_l\, dT_l' \qquad (13.20)$$

Rate of total heat gain by liquid

$$\left(T_g - T_l''\right)U_o P_o\, dx = \dot{m}_l C_l\left(dT_l' - dT_l''\right) \qquad (13.21)$$

Overall heat balance

$$\dot{m}_l C_l\left(T_{l,\,\text{out}} - T_{l,\,\text{in}}\right) = -\dot{m}_g C_g\left(T_{g,\,\text{out}} - T_{g,\,\text{in}}\right) \qquad (13.22)$$

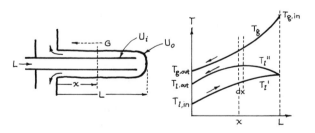

Fig. 13.22. Setting up the heat balance for the bayonet exchanger.

Solving these equations is tedious [see Hurd (1946)], and in terms of

$$X = \frac{T_{g,\,in} - T_{g,\,out}}{T_{l,\,out} - T_{l,\,in}}, \qquad Y = \frac{(T_{g,\,in} + T_{g,\,out}) - (T_{l,\,out} + T_{l,\,in})}{(T_{l,\,out} - T_{l,\,in})},$$

$$Z = \frac{U_i A_i}{U_o A_o} \qquad (13.23)$$

gives

$$\dot{q} = U_o A_o \overline{\Delta T}, \qquad \text{where } \overline{\Delta T} = \frac{(T_{l,\,out} - T_{l,\,in})\left[(X + 1)^2 + 4Z\right]^{1/2}}{\ln \dfrac{Y + \left[(X + 1)^2 + 4Z\right]^{1/2}}{Y - \left[(X + 1)^2 + 4Z\right]^{1/2}}} \qquad (13.24)$$

This equation holds for arrangement (a), and also for arrangement (d). For arrangements (b) and (c), a similar analysis gives

$$\dot{q} = U_o A_o \overline{\Delta T}, \qquad \text{where } \overline{\Delta T} = \frac{(T_{l,\,out} - T_{l,\,in})\left[(X - 1)^2 + 4Z\right]^{1/2}}{\ln \dfrac{Y + \left[(X - 1)^2 + 4Z\right]^{1/2}}{Y - \left[(X - 1)^2 + 4Z\right]^{1/2}}}$$

$$(13.25)$$

Arrangements (e) and (f) represent boiling or condensation or crossflow or very high flow rate of fluid outside the bayonet. For these arrangements simply put $X = 0$ in the above expressions to obtain

$$\dot{q} = U_o A_o \overline{\Delta T}, \qquad \text{where } \overline{\Delta T} = \frac{(T_{l,\,out} - T_{l,\,in})(1 + 4Z)^{1/2}}{\ln \dfrac{Y + (1 + 4Z)^{1/2}}{Y - (1 + 4Z)^{1/2}}} \qquad (13.26)$$

In equations (13.24)–(13.26), whenever $Y < [(X \pm 1)^2 + 4Z]^{1/2}$, there is no real solution to $\overline{\Delta T}$. This means that it is impossible to achieve the desired outlet temperature with the bayonet exchanger.

Examining the above expressions shows that arrangements (b) and (c) are superior to arrangements (a) and (d) in that they give a larger $\overline{\Delta T}$; thus, they require a smaller exchanger area for given inlet and outlet conditions.

V. MIXED FLOW L/PLUG FLOW G EXCHANGERS

The heating vat with immersed heat exchanger coils can be idealized by this contacting pattern. Its sketch and Q vs. T diagram are shown in Fig. 13.23. The key here is to recognize that $T_{l,\text{out}}$ represents the liquid temperature contacted by all the gas. The performance equation, equation (13.1), then becomes

$$\dot{q} = -\left[\dot{m}C(T_{\text{out}} - T_{\text{in}})\right]_g = \left[\dot{m}C(T_{\text{out}} - T_{\text{in}})\right]_l = UA\Delta T_{lm} \quad (13.27)$$

where

$$\Delta T_{lm} = \frac{\left(T_{g,\text{in}} - T_{l,\text{out}}\right) - \left(T_{g,\text{out}} - T_{l,\text{out}}\right)}{\ln\dfrac{T_{g,\text{in}} - T_{l,\text{out}}}{T_{g,\text{out}} - T_{l,\text{out}}}} \quad (13.28)$$

from which

$$T_{g,\text{out}} = \frac{\left[(1 - K)\dot{m}_g C_g + K\dot{m}_l C_l\right]T_{g,\text{in}} + (1 - K)\dot{m}_l C_l T_{l,\text{in}}}{(1 - K)\dot{m}_g C_g + \dot{m}_l C_l} \quad (13.29)$$

$$T_{l,\text{out}} = \frac{(1 - K)\dot{m}_g C_g T_{g,\text{in}} + \dot{m}_l C_l T_{l,\text{in}}}{(1 - K)\dot{m}_g C_g + \dot{m}_l C_l} \quad (13.30)$$

and where

$$K = \exp\left[-\left(UA/\dot{m}_g C_g\right)\right] \quad (13.31)$$

Fig. 13.23

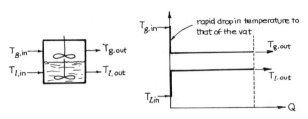

Fig. 13.24

VI. MIXED FLOW L/MIXED FLOW G EXCHANGERS

A stirred vat where the heating fluid swirls about or condenses in a double shell is best approximated by mixed flow/mixed flow, shown in Fig. 13.24. Recognizing that the temperature of the leaving streams represents all the fluid in the vat equation (13.1) becomes

$$\dot{q} = -\left[\dot{m}C(T_{\text{out}} - T_{\text{in}})\right]_g = \left[\dot{m}C(T_{\text{out}} - T_{\text{in}})\right]_l = UA\left(T_{g,\text{out}} - T_{l,\text{out}}\right)$$

$$(13.32)$$

from which

$$T_{g,\text{out}} = \frac{\left(\dfrac{1}{UA} + \dfrac{1}{\dot{m}_l C_l}\right)T_{g,\text{in}} + \dfrac{1}{\dot{m}_g C_g}T_{l,\text{in}}}{\dfrac{1}{\dot{m}_g C_g} + \dfrac{1}{\dot{m}_l C_l} + \dfrac{1}{UA}}$$

$$(13.33)$$

and

$$T_{l,\text{out}} = \frac{\dfrac{1}{\dot{m}_l C_l}T_{g,\text{in}} + \left(\dfrac{1}{UA} + \dfrac{1}{\dot{m}_g C_g}\right)T_{l,\text{in}}}{\dfrac{1}{\dot{m}_g C_g} + \dfrac{1}{\dot{m}_l C_l} + \dfrac{1}{UA}}$$

$$(13.34)$$

VII. HEATING A BATCH OF FLUID

The operations for heating a batch of fluid fall into two broad classes; first, where the liquid is heated within the vat and, second, where a portion of the fluid is removed from the vat, passed though an external heat

Fig. 13.25. Two ways of heating a batch of liquid with a stream of gas.

exchanger, and then returned to the vat. Calling the vat fluid a liquid, and the heating fluid a gas Fig. 13.25 then shows these two types of operations. The particular feature of the external exchanger is that one can have a high liquid velocity through the exchanger tubes, hence high U value. This is a big advantage for viscous, difficult to stir liquids.

With each class of exchanger we can have a variety of contacting patterns. We consider a few simple cases here.

VIII. UNIFORMLY MIXED BATCH *L*/MIXED FLOW *G* EXCHANGERS

This ideal represents a batch of liquid heated by hot gas swirling through the shell of the vessel. We sketch this operation in Fig. 13.26.

Since we have unsteady state operations the three terms of equation (13.1) become:

$$\dot{q} = -\left[\dot{m}C(T_{\text{out}} - T_{\text{in}})\right]_g = W_l C_l \frac{dT_l}{dt} = UA(T_{g,\text{out}} - T_l) \quad (13.35)$$

$$\underbrace{\phantom{-\left[\dot{m}C(T_{\text{out}} - T_{\text{in}})\right]_g}}_{\text{I}} \quad \underbrace{\phantom{W_l C_l \frac{dT_l}{dt}}}_{\text{II}} \quad \underbrace{\phantom{UA(T_{g,\text{out}} - T_l)}}_{\text{III}}$$

Fig. 13.26. Heating a batch of liquid with mixed flow of gas.

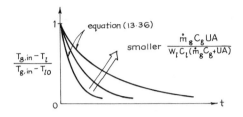

Fig. 13.27. Progress with time in the heating of a batch of liquid.

Here we have three variables to evaluate: t, T_l, and $T_{g,\text{out}}$. First write $T_{g,\text{out}}$ in terms of T_l and t, thus

$$T_{g,\text{out}} = \frac{\dot{m}_g C_g T_{g,\text{in}} + UAT_l}{\dot{m}_g C_g + UA}$$

Eliminating $T_{g,\text{out}}$ in I and II of equation (13.35) and rearranging gives one differential equation in two variables:

$$\frac{dT_l}{dt} = \frac{\dot{m}_g C_g UA\left(T_{g,\text{in}} - T_l\right)}{W_l C_l\left(\dot{m}_g C_g + UA\right)}$$

Separating and integrating from $t = 0$ to t gives:

or

$$\left.\begin{array}{r}\ln\left[\dfrac{T_{g,\text{in}} - T_l}{T_{g,\text{in}} - T_{l0}}\right] = -\dfrac{\dot{m}_g C_g UA}{W_l C_l\left(\dot{m}_g C_g + UA\right)} \cdot t \\[4mm] \dfrac{T_{g,\text{in}} - T_l}{T_{g,\text{in}} - T_{l0}} = \exp\left[\dfrac{\dot{m}_g C_g UA}{W_l C_l\left(\dot{m}_g C_g + UA\right)} \cdot t\right]\end{array}\right\} \qquad (13.36)$$

In graphical representation equation (13.36) is shown in Fig. 13.27. As may be seen, smaller U, A, $\dot{m}_g$, or larger W_l all give a longer heating time.

IX. UNIFORMLY MIXED BATCH L/ISOTHERMAL, MIXED FLOW G (CONDENSATION OR BOILING) EXCHANGERS

The two typical operations represented by this ideal are shown in the sketches of Fig. 13.28. Consider the first of these, in which a batch of liquid is heated by a stream of condensing steam. The sketches of Fig. 13.29 represent this situation. The analysis of the second of these operations of Fig. 13.28 follows in a similar fashion. For this first case the three terms of

Fig. 13.28

equation (13.1) become

$$\dot{q} = \dot{m}_g \lambda_g = W_l C_l \frac{dT_l}{dt} = UA(T_g - T_l) \tag{13.37}$$

Here there are just two variables in II and III. So combining the integrating gives

$$\ln\left[\frac{T_g - T_l}{T_g - T_{l0}}\right] = -\frac{UA}{W_l C_l} \cdot t \quad \text{or} \quad \frac{T_g - T_l}{T_g - T_{l0}} = \exp\left[-\frac{UA}{W_l C_l} \cdot t\right]$$

$$\tag{13.38}$$

The quantity of liquid condensed W_g until any time t is given by a heat balance. Thus

$$W_g \lambda_g = W_l C_l (T_l - T_{l0}), \quad \text{where} \quad \begin{cases} \lambda_g > 0 \text{ for boiling} \\ \lambda_g < 0 \text{ for condensation} \end{cases}$$

$$\tag{13.39}$$

Fig. 13.29. Heating a batch of liquid with condensing vapor.

The temperature–time diagram for this case is very much like the figure for the previous case, Section VIII of this chapter.

X. UNIFORMLY MIXED BATCH L/PLUG FLOW G EXCHANGERS

The sketches representing this operation are shown in Fig. 13.30. If we assume that the temperature of the batch of liquid T_l changes but little during the passage of an element of gas through the heating coils, then equation (13.1) becomes

$$\dot{q} = \underbrace{\dot{m}_g C_g \left(T_{g,\,in} - T_{g,\,out} \right)}_{I} = \underbrace{W_l C_l \frac{dT_l}{dt}}_{II} = \underbrace{UA\Delta T_{lm}}_{III} \qquad (13.40)$$

where

$$\Delta T_{lm} = \frac{\left(T_{g,\,in} - T_l \right) - \left(T_{g,\,out} - T_l \right)}{\ln \dfrac{T_{g,\,in} - T_l}{T_{g,\,out} - T_l}} \qquad (13.41)$$

Writing $T_{g,\,out}$ explicitly with II and III, eliminating it in II, and then solving the differential equation of I and II gives, after considerable manipulation,

$$\frac{T_{g,\,in} - T_l}{T_{g,\,in} - T_{l0}} = \exp\left[-\frac{\dot{m}_g C_g (1 - K)}{W_l C_l} \cdot t \right] \qquad (13.42)$$

Fig. 13.30. Heating a batch of liquid with plug flow of hot gas.

where

$$K = \exp\left[-\frac{UA}{\dot{m}_g C_g} \right] \qquad (13.43)$$

The temperature of exit gas when the liquid is at temperature T_l is then found to be

$$T_{g,\text{out}} = T_l + K\left(T_{g,\text{in}} - T_l\right) \qquad (13.44)$$

Again, the graphical representation is similar to that of Section VIII of this chapter.

XI. UNIFORMLY MIXED BATCH *L*/EXTERNAL EXCHANGER WITH ISOTHERMAL *G*

For this arrangement the sketch with nomenclature is shown in Fig. 13.31. A heat balance and rate expression for exchanger and vat gives:

$$\dot{q} = m_l C_l (T_l' - T_l) = W_l C_l \frac{dT_l}{dt} = UA\Delta T_{lm} \qquad (13.45)$$

I	II	III
Heat entering the vat by recirculating liquid	Heat accumulation in the vat	Transfer rate in the external exchanger

where, for the exchanger,

$$\Delta T_{lm} = \frac{\left(T_g - T_l\right) - \left(T_g - T_l'\right)}{\ln \dfrac{T_g - T_l}{T_g - T_l'}} \qquad (13.46)$$

Fig. 13.31. Heating (or cooling) a batch of liquid by means of an external exchanger using a flow of condensing gas (or boiling liquid).

Combining II and III gives T_l', replacing in I, and then solving I and II gives:

$$\frac{T_g - T_l}{T_g - T_{l0}} = \exp\left[-\frac{\dot{m}_l(1 - K)}{W_l} \cdot t\right]$$ (13.47)

where

$$K = \exp\left[-\frac{UA}{\dot{m}_l C_l}\right]$$ (13.48)

XII. UNIFORMLY MIXED BATCH L/EXTERNAL SHELL AND TUBE EXCHANGER

Here the nomenclature is as shown in Fig. 13.32. The heat balances then become

$$\dot{q} = \underbrace{\dot{m}_l C_l(T_l' - T_l)}_{\substack{\text{I}\\ \text{Heat}\\ \text{added}\\ \text{to vat}}} = \underbrace{W_l C_l \frac{dT_l}{dt}}_{\substack{\text{II}\\ \text{Accumulation}\\ \text{in vat}}} = \underbrace{\dot{m}_g C_g(T_{g,\,\text{out}} - T_{g,\,\text{in}})}_{\substack{\text{III}\\ \text{Heat loss}\\ \text{by gas}}} = \underbrace{UA\Delta T_{lm}\mathscr{F}}_{\substack{\text{IV}\\ \text{Transfer}\\ \text{rate}}}$$

(13.49)

where

$$\Delta T_{lm} = \frac{(T_{g,\,\text{in}} - T_l') - (T_{g,\,\text{out}} - T_l)}{\ln \dfrac{T_{g,\,\text{in}} - T_l'}{T_{g,\,\text{out}} - T_l}}$$ (13.50)

Fig. 13.32. Heating a batch of liquid by means of an external multipass shell and tube exchanger.

and where $\mathscr{F}$ is the correction factor for multipass shell and tube exchangers.

These equations contain four variables: T_l, T_l', $T_{g,\text{out}}$, and t. First eliminate two of them, T_l' and $T_{g,\text{out}}$, from ΔT_{lm}. Thus, from I and II, we write

$$T_l' = T_l + \frac{W_l C_l}{\dot{m}_l C_l} \frac{dT_l}{dt} \qquad (13.51)$$

Then from I and III

$$T_{g,\text{out}} = T_{g,\text{in}} - \frac{W_l C_l}{\dot{m}_g C_g} \frac{dT_l}{dt} \qquad (13.52)$$

Replacing equations (13.51) and (13.52) into I and IV and integrating gives, after considerable manipulation,

$$\frac{T_{g,\text{in}} - T_l}{T_{g,\text{in}} - T_{l0}} = \exp\left[-\frac{1-K}{W_l} \left(\frac{\dot{m}_g \dot{m}_l C_g}{\dot{m}_l C_l - K \dot{m}_g C_g} \right) t \right] \qquad (13.53)$$

where

$$K = \exp\left[-UA\mathscr{F} \left(\frac{1}{\dot{m}_g C_g} - \frac{1}{\dot{m}_l C_l} \right) \right] \qquad (13.54)$$

Again, the temperature time curve for this operation is similar in shape to the curve of section VIII of this chapter.

XIII. FINAL COMMENTS

By far, most of the research in heat transfer aims at finding accurate and reliable values for the heat transfer coefficient h for a variety of situations. Often, however, most of the uncertainty in exchanger design comes from an improper representation of the flow and contacting pattern of the fluids.

For example, in design we consider the shell fluid in an exchanger to be in plug flow. In fact it may be far from plug flow, with stagnant regions, recirculating pockets of fluid, and considerable bypassing, as shown in Fig. 13.33.

Fig. 13.33

On the tube side, the fluid has a choice of flow path among the many parallel tubes in the bundle. Slightly different frictional resistance from tube to tube in clean exchangers will lead to different flow rates of fluid, different tube temperatures, and different buildup rates of scale in these tubes. The final result can be a very wide distribution of residence times of fluid in the tubes, hence large deviation from plug flow.

The $\mathscr{F}$ charts, all based on ideal contacting, can thus be too optimistic. This whole question needs to be carefully looked into. Also, in practical design we must try our best to minimize these deviations from plug flow.

Another all but ignored phenomenon, but one which could lead to a substantial error in prediction concerns the transport of heat from hot to cold regions of the exchanger by the metal walls themselves. To illustrate this phenomenon sketch (a) of Fig. 13.34 shows the temperature distribution in an ideal countercurrent exchanger. But heat transport along the metal walls will give a more uniform temperature in the metal, and thus distort the temperature profile in the fluid, somewhat as shown in sketch (b). The overall effect of this phenomenon is to lower the exchanger efficiency $\mathscr{F}$, the extreme being to cause the efficiency of any shell and tube exchanger to approach that of the stirred vat of Section VI, above. Jakob (1957) considers and presents an analysis of this phenomenon.

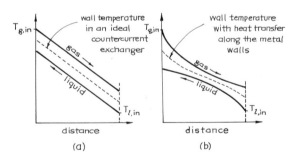

Fig. 13.34. Heat transfer along the metal walls of a recuperator will distort the temperature profile and lower the contact efficiency $\mathscr{F}$.

Fig. 13.35

In this chapter we have only looked at some of the many possible contacting patterns. Equations and $\mathscr{F}$ charts for other contacting patterns may be found in specialized books on the subject.

When heat losses to the surroundings are to be considered the governing expression, equation (13.1), has to be modified somewhat. For example, in the double-pipe exchanger with heat loss to the surroundings, as shown in Fig. 13.35, the basic equations become

$$q_{\text{lost by } g} = q_{\text{gained by } l} + q_{\text{gained by surr}}$$

$$\underset{q_{tx,g \text{ to } l}}{\|} \qquad \underset{q_{tx,g \text{ to surr}}}{\|} \qquad (13.55)$$

Two extra terms
beyond equation (13.1)

instead of equation (13.1).

Finally the overall heat transfer coefficients U have orders of magnitude as shown in Table 13.1.

Table 13.1. Range of U Values in Recuperators[a]

Type of recuperator	U, W/m^2 K
Gas to gas	10–35
Water to water	850–1700
Oil to oil	100–300
Condensing vapor to air (steam pipes in air)	35–90
Condensing steam to boiling organic	280–2300
Condensing steam to boiling water	1700–4500

[a] From Perry (1950). Pages 480–482 give numerous additional values.

REFERENCES AND RELATED READINGS

B. S. Baclic, A simplified formula for cross flow for heat exchanger effectiveness, *ASME J. Heat Transfer* **100**, 746 (1978).

R. A. Bowman, A. C. Mueller, and W. M. Nagle, Mean temperature difference in design, *Trans. ASME* **62**, 283 (1940).

V. Cavesano, ed., *Process Heat Exchange*, Sec. 1, McGraw-Hill, New York (1979). Presents the collective wisdom and practical art of real live designers of recuperators.

N. L. Hurd, Mean temperature difference in the field or bayonet tube, *Ind. Eng. Chem.* **38**, 1266 (1946).

M. Jakob, *Heat Transfer*, Chapter 3, Wiley, New York (1957).

W. A. Kays and A. L. London, *Compact Heat Exchangers*, Second Ed., McGraw-Hill, New York (1964).

D. Q. Kern, *Process Heat Transfer*, McGraw-Hill, New York (1950). An excellent primary reference to recuperators of all types.

A. C. Mueller, in *Handbook of Heat Transfer* (W. M. Roshenow, and J. P., Hartnett, eds.), Sec. 18, McGraw-Hill, New York (1973).

J. H. Perry, *Chemical Engineers' Handbook*, Third Ed., McGraw-Hill, New York (1950).

TEMA, *Standards of Tubular Exchanger Manufacturers Association*, Sixth Ed., Tubular Exchanger Manufacturers Association, Inc., New York (1978).

J. P. Todd and H. B. Ellis, *Applied Heat Transfer*, p. 430, Harper & Row, New York (1982).

R. Turton, C. D. Ferguson, and O. Levenspiel, Performance and design charts for heat exchangers, *J. Heat Transfer*, in press (1984).

EXAMPLE 13.1. EXIT TEMPERATURE FROM A RECUPERATOR

Hot oil (150°C, 1 kg/s, $C_p = 2000$ J/kg K) is to be cooled by cold water (25°C, 1.026 kg/s, $C_p = 4200$ J/kg K) in a countercurrent concentric tube heat exchanger ($A = 4.87$ m^2, $U = 500$ W/m^2 K). Find the temperature of the leaving oil.

Solution

Calling the hot fluid g and the cold fluid l we have from equations (13.7) and (13.8)

$$\frac{T_{g,\text{in}} - T_{g,\text{out}}}{T_{g,\text{in}} - T_{l,\text{in}}} = \frac{1 - \exp\left[-UA\left(\frac{1}{\dot{m}_g C_g} - \frac{1}{\dot{m}_l C_l}\right)\right]}{1 - \frac{\dot{m}_g C_g}{\dot{m}_l C_l}\exp\left[-UA\left(\frac{1}{\dot{m}_g C_g} - \frac{1}{\dot{m}_l C_l}\right)\right]}$$

Replacing values gives

$$\frac{150 - T_{g,\,out}}{150 - 25}$$

$$= \frac{1 - \exp\left[-(500)(4.87)\left(\dfrac{1}{(1)(2000)} - \dfrac{1}{(1.026)(4200)}\right)\right]}{1 - \dfrac{2000}{(1.026)(4200)}\exp\left[-(500)(4.87)\left(\dfrac{1}{(1)(2000)} - \dfrac{1}{(1.026)(4200)}\right)\right]}$$

from which the temperature of the leaving cooled oil is

$$T_{g,\,out} = 71\,°C$$

NOTE: See example 8.10 of Todd and Ellis (1982) for a quite different approach to this problem.

EXAMPLE 13.2. HEATING A BATCH OF LIQUID

Thirty cubic meters of liquid benzene in a storage tank is required to be under pressure and at 150°C for a batch extraction. The storage temperature is 0°C. A pump connected to the tank is able to circulate 4.8 kg/s of benzene to the tube side of a 1-2 shell and tube heat exchanger ($U = 240$ W/m² K, $A = 40$ m²), while heating gas at 2.0 kg/s and 200°C passes through the shell side of the exchanger.

How long will it take to heat the stirred batch of benzene using this external heat exchanger?

Data:

$$\rho_{benzene} = 880 \text{ kg/m}^3, \ C_p(\text{benzene}) = 2000 \text{ J/kg K}$$

$$C_p(\text{heating gas}) = 2400 \text{ J/kg K}$$

Solution

This set-up represents a mixed batch liquid/external shell and tube exchanger which is treated in Section XII. Thus equations (13.53) and (13.54) apply. With values given these equations become:

$$\frac{200-150}{200-0} = \exp\left[-\frac{1-K}{26{,}400}\left\{\frac{2(4.8)(2400)}{4.8(2000)-K(2)(2400)}\right\}t\right] \qquad (i)$$

where

$$K = \exp\left\{-240(40)\,\mathscr{F}\left[\frac{1}{2(2400)}-\frac{1}{4.8(2000)}\right]\right\} \qquad (ii)$$

Now evaluate $\mathscr{F}$ for the 1-2 exchanger. Referring to Fig. 13.5

$$\text{NTU} = \frac{UA}{\dot{m}_l C_l} = \frac{240(40)}{4.8(2000)} = 1$$

and

$$R = \frac{T_{g,\text{in}} - T_{g,\text{out}}}{T_l' - T_l} = \frac{\dot{m}_l C_l}{\dot{m}_g C_g} = \frac{4.8(2000)}{2(2400)} = 2$$

from which

$$\mathscr{F} = 0.76$$

Inserting this value of $\mathscr{F}$ into equation (ii) gives

$$K = 0.4677$$

and inserting this value of K into equation (i) gives

$$t = 21948 \text{ s} \cong 6 \text{ hr } 6 \text{ min}$$

This is the time needed to heat the batch of benzene from 0°C to 150°C.

PROBLEMS ON THE DESIGN OF RECUPERATORS

13.1. Which of the pair of schemes shown on the opposite page do you think is better in the sense that it has a larger $\mathscr{F}$ value. Use qualitative arguments only, give your reasons, and remember that:

- approach to countercurrent plug flow is desired;
- it is more important to improve the end of the exchanger where ΔT is smaller, because it contributes more to the exchanger size.

13.2. The showers in the changing rooms for our gymnasium will need a continuous supply of 1 ton/min of hot water. One way of doing this is by heating city water from 10 to 90°C in a 1-3 pass heat exchanger where water flows through the tubes while waste saturated steam at 1 atm condenses in the shell, leaving as liquid at 100°C. If $U = 1500$ W/m^2 K find:
   (a) the area of exchanger needed, and
   (b) the amount of steam needed (in kg/hr).
   Data:

$$\lambda_{\text{condensation}} = 2.29 \times 10^6 \text{ J/kg}, \quad C_{p,\text{water}} = 4184 \text{ J/kg K}$$

13.3. In the design of fluidized boilers for electricity-generating power stations the heat transfer rate from bed to immersed tubes is an important design parameter. To estimate this let us run the following experiment. Joe Shlemiel

calculates U assuming a well-mixed flow of water in the tube. You feel that this is a poor assumption, and that he should have assumed plug flow.
(a) What did J. S. find with his assumption?
(b) What would you find with your assumption?
(c) What percentage error did J. S. make assuming that you are right?
(d) Which assumption is more reasonable?

13.4. Clean cold water is heated from 10 to 60°C in a countercurrent heat exchanger using an equal flow of hot waste water at 90°C. This 60°C is not hot enough. What must be done to the flow rate of hot water if the cold water is to be heated to 70°C? Assume that U remains unchanged.

13.5. Figure 13.P5 shows a countercurrent concentric pipe heat exchanger ($A = 8$ m^2) for water with a rather unusual piping arrangement. Also shown are all the inlet flow rates and inlet temperatures. If $U = 523$ W/m^2 K evaluate the temperatures T_3 and T_4. Take $C_{p,H_2O} = 4184$ J/kg K at all temperatures.

Hot oil A (300°C) is cooled with fluid B (0°C) in a double-pipe heat exchanger and both fluids leave at 200°C. That is not good enough, so consider the following modifications. Find the outlet temperature of oil A:

13.6. if we add a second identical exchanger in series with the first, in effect doubling the exchanger area.

13.7. if we double the flow rate of coolant. Assume no change in heat transfer coefficient.

13.8. A certain process generates steam at atmospheric pressure and when superheated to 600°C. It is desired to recover the heat content of this steam by
countercurrent heat exchange with cold dry air entering at 0°C.

(a) Construct a plot of Q vs. T for steam at atmospheric pressure from
0°C to 600°C, expressing Q in J/mol.

(b) With the aid of this plot calculate the largest quantity of cold air
which can be heated to 600°C using perfect countercurrent heat exchange
(mol air/mol steam).

(c) What percent of the total heat in the original steam is recovered by
the process of part (b).

(d) Calculate the smallest number of moles of air which by countercurrent flow can extract all the heat from 1 mole of steam entering the
exchanger; in effect, cool this stream down to 0°C.

(e) To what temperature will the air in (d) be heated?

Data: From thermodynamic tables:

$$C_{p,\text{air}} = 29.29 \text{ J/mol K}$$

$$H \text{ (steam, 600°C)} = 3704.7 \times 10^3 \text{ J/kg}$$

$$H \text{ (steam, 100°C)} = 2675.5 \times 10^3 \text{ J/kg}$$

$$H \text{ (water, 100°C)} = 417.5 \times 10^3 \text{ J/kg}$$

$$H \text{ (water, 0°C)} = 0$$

13.9. A water-cooled countercurrent flow exchanger in a refrigeration plant is used
for condensing ammonia. Superheated ammonia enters the condenser at
50°C, condenses at 25°C and liquid NH_3 is then cooled to 6°C before
leaving the exchanger. Cooling water enters at 5°C and leaves at 15°C,
flowing at 20 kg/min. Calculate the condenser area needed if

$$U \text{ (between } NH_3 \text{ vapor and } H_2O) = 60 \text{ W/m}^2 \text{ K}$$

$$U \text{ (between condensing } NH_3 \text{ and } H_2O) = 360 \text{ W/m}^2 \text{ K}$$

$$U \text{ (between liquid } NH_3 \text{ and } H_2O) = 120 \text{ W/m}^2 \text{ K}$$

Data: From thermodynamic tables:

$$H \text{ (}NH_3 \text{ vapor at 50°C)} = 1537.7 \times 10^3 \text{ J/kg}$$

$$H \text{ (saturated } NH_3 \text{ vapor, 25°C)} = 1465.0 \times 10^3 \text{ J/kg}$$

$$H \text{ (boiling } NH_3 \text{ liquid, 25°C)} = 298.8 \times 10^3 \text{ J/kg}$$

$$H \text{ (}NH_3 \text{ liquid at 6°C)} = 208.9 \times 10^3 \text{ J/kg}$$

$$H \text{ (}H_2O \text{ at 5°C)} = 21000 \text{ J/kg}$$

$$H \text{ (}H_2O \text{ at 15°C)} = 63000 \text{ J/kg}$$

NOTE: This question is similar to the one asked in the Professional Engineers examination, given by the State of Oregon in the 1950s, except that they did not give the enthalpies.

13.10. For the end conditions shown below, compare the heat exchanger efficiency of the following contacting schemes with respect to countercurrent plug flow. In effect determine $\mathscr{F}$ for each scheme:

  (a) cocurrent flow;
  (b) 1-2 shell and tube exchanger;
  (c) 2-4 shell and tube exchanger;
  (d) cross flow, unmixed exchanger;
  (e) cross flow, mixed gas, unmixed liquid exchanger;
  (f) cross flow, unmixed gas, mixed liquid exchanger.

13.11. A steam of cold air at 0°C is heated to 800°C in a 1-2 shell and tube heat exchanger using $2\frac{1}{2}$ times as much hot air entering at 1000°C. Which of the following two piping arrangements would result in a smaller exchanger area, to have the hot gas flow through the shell side or through the tube side? Show calculations to support your answer.

13.12. Light lubricating oil at 110°C is cooled with brackish water at 10°C in a 3-6 shell and tube heat exchanger with oil in the shell side. When the exchanger is clean the oil leaves at 35°C and the water leaves at 60°C. With time, however, scale forms in the tubes and the efficiency of heat transfer becomes lower. Today the exchanger is quite dirty and the oil leaves at 50°C. Find $U_{\text{dirty}}/U_{\text{clean}}$ and $\mathscr{F}_{\text{dirty}}/\mathscr{F}_{\text{clean}}$. (Problem prepared by Carol Magnusson.)

An oil (150°C, 10 kg/s, C_p = 3750 J/kg K) is to be cooled to as low a temperature as possible using cooling water (35°C, 22.5 kg/s).

13.13. In a 1-2 shell and tube exchanger (U = 750 W/m² K, A = 100 m²) should oil be in the shell or tube side? Determine the exit oil temperature for these two alternatives.

13.14. We have available both a 1-2 and a 4-8 exchanger. Both have U = 750 W/m² K and A = 100 m². With oil in the shell side which exchanger gives a cooler leaving oil? Evaluate these temperatures.

13.15. Determine the length of bayonet exchangers needed in a 21-tube bundle which is to be used to heat 12.5 ton/hr of cold fluid (C_p = 2 kJ/kg K) from 0 to 40°C by shell fluid (C_p = 4 kJ/kg K) cooling from 120 to 40°C. For the inner tube $d = 3$ cm, $U = 1000$ W/m^2 K. For the outer tube $d = 6$ cm, $U = 500$ W/m^2 K.

(a) Assume that cold fluid enters at A (see Fig. 13.P15).

(b) Assume that cold fluid enters at B (see Fig. 13.P15).

(c) Which routing is better?

13.16. Whales, dolphins and porpoises are able to maintain surprisingly high body temperatures even though they are immersed continuously in very cold water. Since the extremities of these animals (tails, fins, flukes) have a large surface much of the heat loss occurs there.

(a) Now an engineering student designing a dolphin from first principles would probably view the heat loss from a flipper somewhat as shown here:

Let us suppose that blood at 40°C enters a flipper at 0.3 kg/s, feeds the flipper, is cooled somewhat, and then returns to the main part of the body. The dolphin swims in 4°C water, the flipper area is 3 m^2 and $U = 418.4$ W/m^2 K between the flipper and water. At what temperature does blood reenter the main body of the dolphin?

(b) Frankly, an ordinary engineering student would design a lousy dolphin. Let us try to do better; in fact, let us learn from nature, and see whether we can reduce some of the undesirable heat loss by transferring heat from the outgoing warm arterial blood to the cooled venous blood. Such a scheme is idealized above. Assume for this internal exchanger that $A = 2.4$ m^2 and $U = 523$ W/m^2 K. With this extra exchanger find the temperature of the blood returning to the main part of the body; and, in addition, the fraction of original heat loss which is saved.

Data: Approximate the properties of blood by water.

A continuous stream of 6 tons/hr of cold water is to be heated from 0°C to 95°C using a stream of hot steam at 260°C and 1 atm which cools, condenses and leaves as liquid at very close to 100°C. Find the exchanger area needed and the hourly steam rate:

13.17. for countercurrent contacting.

13.18. if the water is heated in a big stirred vat containing a long immersed coil for the cooling and condensing steam.

13.19. if the steam bubbles and condenses into its own condensate in a big vat, while the water to be heated passes through a long coiled tube immersed in the hot condensate.

13.20. if steam bubbles and condenses in its condensate in a vat, while the water to be heated passes in mixed flow through the shell of the vat.

Data: Use the following approximate values:

$$H \text{ (water 0°C)} = 0$$

$$H \text{ (water, 95°C)} = 400 \times 10^3 \text{ J/kg}$$

$$H \text{ (water, 100°C)} = 420 \times 10^3 \text{ J/kg}$$

$$H \text{ (steam, 100°C)} = 2680 \times 10^3 \text{ J/kg}$$

$$H \text{ (steam, 260°C)} = 3000 \times 10^3 \text{ J/kg}$$

$$U_{\text{condensing section}} = 2000 \text{ W/m}^2 \text{ K}$$

$$U_{\text{steam-water}} = 500 \text{ W/m}^2 \text{ K}$$

$$U_{\text{water-water section}} = 1275 \text{ W/m}^2 \text{ K}$$

13.21. Repeat example 13.2 with the following changes: the liquid flow rate through the exchanger is tripled, the gas flow rate is quadrupled, in which case the overall heat transfer coefficient doubles.

13.22. Repeat example 13.2 with just one change: the hot gas used to heat the benzene enters the exchanger at saturated conditions at 200°C, condenses there and leaves at 200°C. In this situation U becomes 600 W/m² K. Now how long does it take to heat the benzene?

My mountain cabin sits close to a mineral spring—smelly but hot. I can't use this water directly for baths but I can use it to heat up fresh cold water in my unique bath tub—an old discarded well-insulated jacketed kettle from a plastic factory. Hot water at 67.5°C enters the jacket at 34.4 liters/min, swirls about, and leaves. The tub is filled with 860 liters of fresh cold water for a family bath.

13.23. Find the time needed to heat this tubful of cold water from 5°C to a comfortable toasty 55°C.

Data: Between hot and cold water: $U = 400$ W/m² K, $A = 3$ m². The outside of the vat is well insulated so assume that $U = 0$ to surroundings.

13.24. Find the time needed to heat this tubful of cold water from 5 to 55°C if the insulation is torn off to expose an attractive copper surface.

Data: Without insulation: $U = 240$ W/m² K, $A = 5$ m² to the surroundings which are at 9°C.

13.25. We plan to cool one-ton batches of oil ($T = 300°C$, $C_p = 3000$ J/kg K) down to 120°C in 2 hr in a stirred vat by pumping nearly boiling water through coiled tubes which are immersed in the hot oil, and withdrawing steam at roughly 100°C.

(a) How much exchanger surface do we need?

(b) How much steam is generated during this 2-hr process?

Data: For the cooling coil

$$U = 1000 \text{ W/m}^2 \text{ K}$$

To boil water

$$\lambda = 2.29 \times 10^6 \text{ J/kg}$$

13.26. A one-ton batch of hot oil ($C_p = 3000$ J/kg) is to be cooled from 300°C to 120°C in a stirred vat by passing cold water (0°C, 0.5 kg/s) in plug flow through heat exchanger coils immersed in the hot oil.

(a) Find how long it will take to cool the oil.

(b) What is the temperature of the water leaving the coils at the beginning and at the end of the run?

Data: For the exchanger coil

$$U = 1000 \text{ W/m}^2 \text{ K}, \quad A = 1.5 \text{ m}^2$$

For water

$$C_p = 4184 \text{ J/kg K}$$

Chapter 14   DIRECT-CONTACT
GAS–SOLID NONSTORING
EXCHANGERS

In this chapter, we first take up the fluidized bed heat exchanger, and then various other devices such as moving grates and moving bed exchangers.

I. FLUIDIZED BED HEAT EXCHANGERS—PRELIMINARY CONSIDERATIONS

Consider the fluidized bed of Fig. 14.1 wherein a stream of cold particles and a stream of hot gas enter, exchange heat, and then leave. Let us examine some of the properties of this stream of solids. From these findings we will be in a position to choose a reasonable set of assumptions to represent the gas–solid fluidized bed exchanger.

1. ΔT *within single particles*. Suppose a particle, initially at T_{s0}, suddenly finds itself in surroundings at T_g. Let us ask how long it would take for the particle to approach the temperature of its surroundings. This we call the thermal relaxation time of the particle.

As an estimate of the relaxation time let us determine the time needed for a 90% approach to equilibrium conditions, of particles of various sizes

cold
solids
in

gas out

suspended, vigorously
mixed fine solids

solids out

hot gas in

Fig. 14.1

moving through 100°C air at 1 m/s. Following the calculation method of Example 11.1 we obtain the values of Table 14.1. Clearly, for particles about d_p = 1 mm or smaller the relaxation time is very small; certainly an order of magnitude smaller than the mean residence time of solids in the vessel.

2. ΔT *among particles*. Because of the large heat capacity of solids (about 1000 times that of gas, per unit of volume) and because of the rapid circulation of solids in the fluidized bed we can assume that the particles are uniform in temperature everywhere in the bed.

3. ΔT *between particles and exit gases*. To get a rough order of magnitude estimate of this ΔT, assume plug flow of gas and well-mixed solids in the bed. Referring to Fig. 14.2, a heat balance across the slice of bed of thickness dx then gives

$$-d\dot{q} = \dot{m}_g C_g \, dT_g = \rho_g u_0 A_t C_g \, dT_g = hA_{tx}\left(T_g - T_s\right) dx$$

Table 14.1. Thermal Relaxation Times for Spherical Particles Passing through Air at 1 m/s

	Relaxation time, t			
	$d_p = 100\ \mu m$	$d_p = 1$ mm	$d_p = 10$ mm	$d_p = 100$ mm
Copper	130 ms[a]	7.5 s[a]	320 s[a]	190 min[a]
	(0.03 ms)[c]	(0.003 s)[c]	(0.3 s)[c]	(0.5 min)[c]
Iron	140 ms[a]	7.8 s[a]	330 s[a]	190 min[a]
Stainless steel	140 ms[a]	7.8 s[a]	330 s[a]	190 min[a]
Glass	64 ms[a]	3.7 s[a]	150 s[a]	110 min
Sand	80 ms[a]	4.5 s[a]	240 s	160 min
PVC plastic	56 ms[a]	3.8 s	170 s	150 min
Wood	40 ms[a]	3.4 s	160 s	130 min
Ice	80 ms[a]	4.5 s[a]	190 s[a]	110 min[a]
Na-K (56% Na)[b]	39 ms[a]	2.2 s[a]	92 s[a]	54 min[a]
Water[b]	160 ms[a]	9.2 s[a]	440 s	290 min

[a]Film resistance controls, Bi < 0.1. All other values in the table fall in the mixed control (particle conduction and film resistance) regime.
[b]Assume no circulation in the particle. In the presence of circulation the relaxation time is much reduced.
[c]( ) Values assuming no film resistance. These values are greatly in error.

Fig. 14.2. Development of the heat balance between gas and solids in a fluidized bed.

Replacing values, rearranging and integrating gives

$$\ln\frac{T_{g,\,\text{in}} - T_s}{T_{g,\,\text{out}} - T_s} = \frac{haL}{\rho_g u_0 C_g} = \frac{\text{Nu}_p}{\text{Pr} \cdot \text{Re}_p} \cdot \frac{6(1 - \varepsilon_f)L}{d_p}$$

Now for a 95% temperature approach of gas to the temperature of the particles

$$\ln\frac{T_{g,\,\text{in}} - T_s}{T_{g,\,\text{out}} - T_s} = \ln 20 \geq 3$$

and combining the above two expressions gives

$$\frac{L}{d_p} \geq 0.5\frac{\text{Pr} \cdot \text{Re}_p}{\text{Nu}_p(1 - \varepsilon_f)}$$

Typically for 0.5-mm particles we find

$$\frac{L}{d} > 25.6 \quad \text{or} \quad L > 1.28 \text{ cm}$$

So, for beds deeper than 1 cm, we can reasonably assume that $T_{g,\text{out}} \cong T_s$, and consequently, for beds deeper than this, we can take $h \cong \infty$ between gas and solid, or more properly, $(h_{g-s}a) \cong \infty$.

4. To *summarize* the findings of this preliminary analysis: All particles in a fluidized bed are isothermal and at the same temperature. In addition, gas leaves at the temperature of the bed solids and $(h_{g-s}a) \cong \infty$. These then are the assumptions which we will use to characterize the behavior of gas–solid fluidized bed heat exchangers.

II. MIXED FLOW G/MIXED FLOW S, OR SINGLE-STAGE FLUIDIZED BED EXCHANGERS

Consider the fluidized bed exchanger of Fig. 14.3. A heat balance over the whole unit gives

$$-\dot{m}_g C_g \left(T_1 - T_{g,\,in} \right) = \dot{m}_s C_s \left(T_1 - T_{s,\,in} \right) \tag{14.1}$$

Note that no rate equation is used here. This is because we took $(h_{g-s}a) = \infty$. So on rearranging we find

$$T_1 = \frac{T_{s,\,in} + \phi T_{g,\,in}}{1 + \phi} \tag{14.2}$$

where ϕ is the heat flow ratio of the two flowing streams, defined as

$$\phi = \frac{\dot{m}_g C_g}{\dot{m}_s C_s} = -\frac{\Delta T_s}{\Delta T_g} = \frac{W/\dot{m}_s}{\left(1 - \varepsilon_f\right)\left(L/u_0\right)} \cdot \frac{\rho_g C_g}{\rho_s C_s} \tag{14.3}$$

The efficiencies of heat utilization are then, from equation (14.2)

$$\left.\begin{array}{l} \eta_g = \dfrac{\Delta T_g}{\Delta T_{g,\,max}} = \dfrac{T_{g,\,in} - T_1}{T_{g,\,in} - T_{s,\,in}} = \dfrac{1}{1 + \phi} \\[3mm] \eta_s = \dfrac{\Delta T_s}{\Delta T_{s,\,max}} = \dfrac{T_1 - T_{s,\,in}}{T_{g,\,in} - T_{s,\,in}} = \dfrac{\phi}{1 + \phi} = 1 - \eta_g \end{array}\right\} \tag{14.4}$$

Efficiencies of heat utilization are always low in single fluidized bed heat exchangers. For example if $\eta = 0.7$ for the gas, it will be 0.3 for the solid (always sum up to 1). To raise the efficiency we multistage, either crossflow or counterflow. We consider these next.

note that the exit gas, the exit solids and the bed solids are all at the same temperature, T_1

Fig. 14.3. A single-stage fluidized bed heat exchanger with through-flow of gas and solids.

III. COUNTERFLOW STAGEWISE FLUIDIZED BED EXCHANGERS

Consider N beds of equal size, as shown in Fig. 14.4. A heat balance around each bed gives

$$
\left.\begin{array}{ll}
\text{Bed 1:} & T_1 - T_{s,\,\text{in}} = \phi(T_2 - T_1) \\[2ex]
\text{Bed 2:} & T_2 - T_1 = \phi(T_3 - T_2) \\[2ex]
\quad \vdots \qquad \qquad \vdots \qquad \qquad \vdots \\[2ex]
\text{Bed } N: & T_N - T_{N-1} = \phi(T_{g,\,\text{in}} - T_N)
\end{array}\right\} \tag{14.5}
$$

Combining equations (14.5) and eliminating intermediate temperatures gives

$$
\left.\begin{array}{l}
\eta_g = \dfrac{T_{g,\,\text{in}} - T_1}{T_{g,\,\text{in}} - T_{s,\,\text{in}}} = \dfrac{\displaystyle\sum_{N=1}^{N} \phi^N}{\phi\left(1 + \displaystyle\sum_{N=1}^{N} \phi^N\right)} \\[5ex]
\eta_s = \dfrac{T_N - T_{s,\,\text{in}}}{T_{g,\,\text{in}} - T_{s,\,\text{in}}} = \dfrac{\displaystyle\sum_{N=1}^{N} \phi^N}{1 + \displaystyle\sum_{N=1}^{N} \phi^N} = \phi\eta_g
\end{array}\right\} \tag{14.6}
$$

Fig. 14.4. A multistage counterflow fluidized bed heat exchanger.

Adjusting flow rates so that $\phi = 1$ (if one stream loses 100°, the other gains 100°) gives

$$\eta_g = \eta_s = \frac{N}{N+1} \qquad (14.7)$$

For a large number of stages, $N \to \infty$, countercurrent plug flow is approached and

$$\left.\begin{array}{l}
\text{for equal heat flows, or } \phi = 1: \quad \eta_g = \eta_s = 1 \\[2mm]
\text{for excess of solids, or } \phi < 1: \quad \eta_g = 1 \text{ and } \eta_s = \phi \\[2mm]
\text{for excess of gas, or } \phi > 1: \quad \eta_g = \dfrac{1}{\phi} \text{ and } \eta_s = 1
\end{array}\right\} \qquad (14.8)$$

IV. CROSSFLOW STAGEWISE FLUIDIZED BED EXCHANGERS

Consider N beds of equal size having the same gas flow through each stage, as shown in Fig. 14.5. A heat balance about each bed gives

$$\left.\begin{array}{ll}
\text{Bed 1:} & T_1 - T_{s,\,\text{in}} = \phi'\left(T_{g,\,\text{in}} - T_1\right) \\[2mm]
\text{Bed 2:} & T_2 - T_1 = \phi'\left(T_{g,\,\text{in}} - T_2\right) \\[2mm]
\quad\vdots & \quad\vdots \qquad\qquad \vdots \\[2mm]
\text{Bed } N: & T_N - T_{N-1} = \phi'\left(T_{g,\,\text{in}} - T_N\right)
\end{array}\right\} \qquad (14.9)$$

Fig. 14.5. A multistage crossflow fluidized bed heat exchanger.

where ϕ' is based on the heat flow through each stage, or

$$\phi' = \frac{(\dot{m}_g/N) \cdot C_g}{\dot{m}_s C_s} = \frac{\phi_{\text{countercurrent}}}{N} \tag{14.10}$$

Now with equal gas flows through each of the N stages we have

$$T_{g,\text{out}} = \frac{T_1 + T_2 + \cdots + T_N}{N} \tag{14.11}$$

Combining equations (14.9) and (14.11) gives the efficiencies of operation as

$$\left.\begin{aligned}
\eta_g &= \frac{\Delta T_g}{\Delta T_{g,\max}} = \frac{T_{g,\text{in}} - T_{g,\text{out}}}{T_{g,\text{in}} - T_{s,\text{in}}} = \frac{1}{N\phi'}\left[1 - \frac{1}{(1+\phi')^N}\right] \\[2ex]
\eta_s &= \frac{\Delta T_s}{\Delta T_{s,\max}} = \frac{T_N - T_{s,\text{in}}}{T_{g,\text{in}} - T_{s,\text{in}}} = \left[1 - \frac{1}{(1+\phi')^N}\right] = N\phi'\eta_g
\end{aligned}\right\} \tag{14.12}$$

Comparing crossflow and counterflow operations shows that for any number of stages N, counterflow has the advantage of being more efficient thermally. However, counterflow has the drawbacks of higher pressure drop, more hydraulic problems, especially with downcomers, and more mechanical design complications.

V. COUNTERCURRENT PLUG FLOW EXCHANGERS

This contacting pattern, see Fig. 14.6, approximates moving bed and shaft kiln operations, and its Q vs. T diagram is shown in Fig. 14.7.

With a stream of small particles the surface area of contact is large and the particle relaxation time is short compared to the time of passage of particles through the exchanger (see Section I of this chapter). The behavior of such systems is equivalent to having a very large $(h_{g-s}a)$ value, approaching infinity. This means that all the heat exchange takes place in a very narrow zone of the exchanger. Consequently, the temperature vs. distance diagram for this operation is as sketched in Fig. 14.8. Note that for

Fig. 14.6. Countercurrent gas–solid contacting.

Fig. 14.7. Q vs. T diagram for countercurrent plug flow of gas and solid wherein $(h_{g-s}a) = \infty$.

close to equal heat flow rates, or $\phi \cong 1$, the location of the temperature front is uncertain. With a slight excess of solids ($\phi = 1 - \varepsilon$) the front slowly migrates to the gas inlet, point A in Fig. 14.8. With a slight excess of gas it slowly migrates to the solid inlet, point B in Fig. 14.8.

Compare the corresponding sketches of Fig. 14.7 and Fig. 14.8 to satisfy yourself that they make sense.

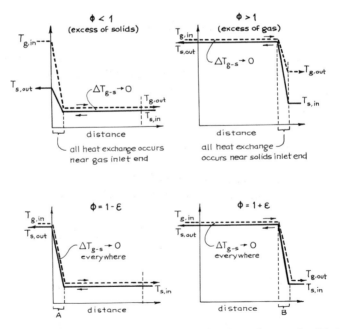

Fig. 14.8. Temperature profiles for countercurrent plug flow of gas and solid. For small particles, all heat exchange takes place near one or other end of the exchanger.

A heat balance over the whole exchanger gives

$$
\left.
\begin{array}{ll}
\text{for equal heat flows, or } \phi = 1: & \eta_g = \eta_s = 1 \\[1.5em]
\text{for an excess of solids, or } \phi < 1: & \eta_g = 1 \text{ and } \eta_s = \phi \\[1.5em]
\text{for an excess of gas, or } \phi > 1: & \eta_g = \dfrac{1}{\phi} \text{ and } \eta_s = 1
\end{array}
\right\} \quad (14.13)
$$

Comments

1. These expressions show that the stream with lower heat flow achieves 100% heat utilization; the stream in excess does not.

2. For very small particles the heat exchange zone is very narrow, and the temperature of gas and solid are very close to each other nearly everywhere. For larger particles the heat exchange zone broadens, and for very large particles one may have to account for nonuniformity in temperature within particles. This type of problem can be treated by a direct extension of the analysis of Chapter 15.

VI. CROSSFLOW OF GAS AND SOLIDS

The way we analyze this contacting pattern depends on the temperature distribution of the flowing solids. There are three extremes that we may want to consider. In all of these cases we assume a relatively short relaxation time, or $(h_{g-s}a) \to \infty$.

A. Well-Mixed Solids / Plug Flow Gas

The analysis is analogous to that in Section 13.VI with UA replaced by ha. And for $ha \to \infty$ equations (13.21) and (13.22) reduce to

$$
\left.
\begin{array}{l}
T_{g,\text{out}} = T_{s,\text{out}} = \dfrac{\dot{m}_g C_g T_{g,\text{in}} + \dot{m}_s C_s T_{s,\text{in}}}{\dot{m}_g C_g + \dot{m}_s C_s} \\[1.5em]
\eta_g = \dfrac{1}{\phi + 1} \\[1.5em]
\eta_s = \dfrac{\phi}{\phi + 1}
\end{array}
\right\} \quad (14.14)
$$

Fluidized beds of large particles approximate this extreme.

B. Solids Mixed Laterally but Unmixed Along Flow Path / Plug Flow Gas

This extreme is representative of a thin stream of solids being contacted crosswise by gas, as shown in Fig. 14.9.

This contacting pattern is identical to the crossflow stagewise fluidized bed exchanger, treated in Section IV, but with an infinite number of stages. So letting $N \to \infty$ in equations (14.12) gives

$$\left.\begin{array}{l} \eta_g = \dfrac{\Delta T_g}{\Delta T_{g,\,\mathrm{max}}} = \dfrac{1}{\phi}[1 - e^{-\phi}] \\[2em] \eta_s = \dfrac{\Delta T_s}{\Delta T_{s,\,\mathrm{max}}} = 1 - e^{-\phi} \end{array}\right\} \qquad (14.15)$$

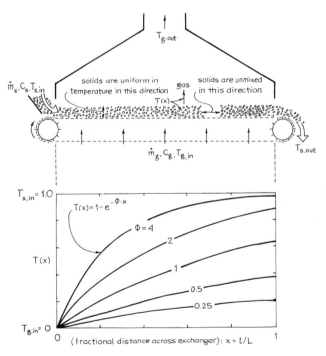

Fig. 14.9. Crossflow exchanger with a thin layer of solids, and its temperature distribution at various gas–solid flow ratios, ϕ.

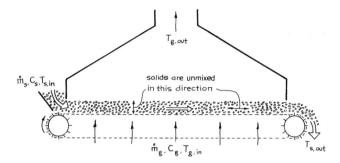

Fig. 14.10. Crossflow exchanger with a thick layer of solids (unmixed vertically).

C. Solids Unmixed / Plug Flow Gas

This extreme represents a thick stream of solids contacted crosswise by gas, as shown in Fig. 14.10. For plug flow of both hot gas and cold solids a sharp temperature front exists as shown in Fig. 14.11, thus the efficiency of heat utilization is:

$$\left.\begin{array}{l} \text{for excess of solids, } \phi < 1: \quad \eta_g = 1 \text{ and } \eta_s = \phi \\[2em] \text{for excess of gas, } \phi > 1: \quad \eta_g = \frac{1}{\phi} \text{ and } \eta_s = 1 \end{array}\right\} \quad (14.16)$$

VII. COMMENTS

The whole treatment of this chapter assumes a short temperature relaxation time for the solids. This assumption is quite reasonable for fluidized beds of fine particles with their very large surface-to-volume ratios.

Fig. 14.11. Temperature distribution in a crossflow exchanger with a thick layer of solids and $(h_{g-s}a) = \infty$.

cold solids ⌐   ⌐ band of warming solids

cold
solids ⟹

hot
solids

Fig. 14.12. Crossflow heating of a
stream of large particles for which
$(h_{g-s}a)$ is small.

hot gas

However, conveyor belts and moving beds of solids often are used to heat or cool streams of large particles. These do not respond rapidly to temperature changes, as is clearly shown in Table 14.1. In addition, gas dispersion and backmixing in these large particle systems can result in serious deviations from plug flow for the gas. Both these factors cause a blurring and broadening of the temperature front in these exchangers, as sketched in Fig. 14.12.

The extent of this broadening depends on the temperature relaxation time of the solids and the extent of gas dispersion compared to the residence time of gas and solid in the gas–solid exchanger. This effect is strongly affected by an increase in particle size. The analysis of this situation is not easy but can be developed following the analysis of Chapter 15.

RELATED READINGS

D. Kunii and O. Levenspiel, *Fluidization Engineering*, Chapter 12, Krieger, Melbourne, FL (1979).

EXAMPLE 14.1. COUNTERFLOW MULTISTAGE FLUIDIZED BED EXCHANGER

We plan to cool a continuous stream of hot solids from 820°C to 220°C with cold gas at 20°C in staged counterflow fluidized beds.

(a) Determine the number of stages needed; see drawing below.
(b) Find the temperature of the flowing streams in the exchanger.

Data. Adjust the flow rates of gas and solid so as to obtain the same thermal utilization for the two streams, or $\eta_g = \eta_s$.

Solution

(a) From the information for the stream of solids:

$$\eta_s = \frac{\Delta T_s}{\Delta T_{s,\,\max}} = \frac{220 - 820}{20 - 820} = 0.75$$

For equal thermal utilization, $\eta_s = \eta_g$, equation (14.7) gives

$$0.75 = \frac{N}{N + 1}$$

thus, the number of stages required is

$$N = 3$$

(b) For the gas stream

$$\eta_g = 0.75 = \frac{\Delta T_g}{\Delta T_{g,\,\max}} = \frac{T_1 - 20}{820 - 20}$$

hence the exit gas temperature

$$T_1 = 620°C.$$

Equation (14.5) then gives the temperature distribution within the exchanger. Thus, with $\phi = 1$,

$$620 - 820 = 1\,(T_2 - 620) \qquad \text{or } T_2 = 420°C$$
$$420 - 620 = 1\,(T_3 - 420) \qquad \text{or } T_3 = 220°C$$

Thus, the final sketch is shown below.

EXAMPLE 14.2. CROSSFLOW MULTISTAGE FLUIDIZED BED EXCHANGER

Repeat Example 14.1 with just one change—use a crossflow rather than a counterflow exchanger; see below.

Solution

For equal thermal utilization we have, as with Example 14.1,

$$\eta_s = 0.75, \qquad \eta_g = 0.75, \qquad \frac{\Delta T_g}{\Delta T_s} = 1, \qquad \frac{\dot{m}_g C_g}{\dot{m}_s C_s} = 1$$

Now since ϕ' refers to the flow through each stage

$$\phi' = \frac{(\dot{m}_g/N) C_g}{\dot{m}_s C_s} = \frac{1}{N} \cdot \frac{\dot{m}_g C_g}{\dot{m}_s C_s} = \frac{1}{N}$$

Then equation (14.12) gives

$$\eta_s = 0.75 = 1 - \frac{1}{(1 + \phi')^N} = 1 - \frac{1}{[1 + (1/N)]^N}$$

or

$$\left(\frac{N+1}{N}\right)^N = 4$$

Solve by trial and error

Guess N	$\left(\dfrac{N+1}{N}\right)^N$
1	2
3	2.37
9	2.58
99	2.70

The above progression does not seem to be able to reach the desired value of "4". We verify this suspicion by examining the limit, or

$$\lim_{N \to \infty} \left(\frac{N+1}{N} \right)^N = \lim_{N \to \infty} \left(1 + \frac{1}{N} \right)^N = e = 2.718$$

This shows that one cannot get a solution to equation (ii). This means that

> It is impossible to get the required 75% heat utilization in any crossflow exchanger, even though this can be done in 3-stage counterflow exchangers.

PROBLEMS ON DIRECT-CONTACT GAS–SOLID NONSTORING EXCHANGERS

14.1. A stream of fine solids is to be cooled from 820°C to 100°C by cold gas heating from 20°C to 500°C in a counterflow multistage fluidized bed exchanger. How many ideal stages are needed?

14.2. A crossflow multistage fluidized bed exchanger is to cool a stream of fine solids from 820°C to 220°C by contact with gas which heats from 20°C to 420°C. How many ideal stages are needed?

14.3. Suppose the equipment for example 14.1 is built and is working smoothly. Then one day we are told that the temperature of the incoming solids will be raised to 1020°C . We still want the solids to leave at 220°C, and we don't want to modify the equipment, add stages, etc. What do you suggest we do?

14.4. We wish to heat a stream of solids (1 ton/hr, $C_p = 1000$ J/kg K) from 0°C to 800°C in a 4-stage direct-contact crossflow fluidized bed exchanger using hot air at 1000°C. What is the required volumetric flow rate of air?

Data: For incoming air: $\pi = 120$ kPa, $C_p = 32$ J/mol K.

Our factory needed to cool a continuous stream of 880°C solids with a stream of 40°C air. For this purpose we built a single-stage fluidized bed exchanger as shown in (a), and for the flow rates of our process we find that the solids leave at 320°C. This is not bad but it is not good enough. Let us try to improve the operations.

(a)   (b)   (c)

14.5. One thought is to put a vertical baffle right down the middle of the unit as shown in (b). Keeping all flow rates unchanged, find the temperature of solids leaving the exchanger.

14.6. Another thought is to insert a horizontal baffle in the exchanger, as shown in (c). Again, with flow rates unchanged, find the temperature of the leaving ᐟsolids.

14.7. At present a stream of hot gas (1000°C) contacts a stream of cold solids (0°C) in a single-stage fluidized bed exchanger. The stream of solids leaves at 500°C. This is not good enough so let us add baffles so as to divide the exchanger into four equal parts, as shown below, keeping all flows unchanged. With this modification find the temperature of the leaving solids and the leaving gas.

14.8. Hot gas at 1000°C enters a three-stage direct-contact crossflow fluidized bed exchanger and contacts a stream of solids which enter at 0°C. The solids leave at 512°C. We now add vertical baffles so as to divide each stage into two, thus ending up with six crossflow stages.

(a) By making this change what happens to the temperature of the leaving solids.

(b) How does this change affect the contacting efficiency for the two streams. To answer calculate the contacting efficiency before and after the change.

14.9. Consider a gas–solid direct-contact fluidized bed heat exchanger. If a four-stage crossflow exchanger can heat solids from 20°C to 820°C using hot entering gas at 1020°C how many stages counterflow (with the same entering and leaving temperatures for gas and solids) can do the same job?

14.10. Consider a gas solid direct contact fluidized bed heat exchanger. If a five-stage crossflow exchanger is needed to heat the solids from 0° to 600°C using hot gas at 1000°C, how many stages counterflow can do the same job if the inlet gas and solids flow rates and temperatures remain unchanged?

14.11. *Moving bed contactor.* 10 tons/hr of solids ($C_p = 800$ J/kg K, $T_{s,in} = 1000$ K) enter the top of a vertical pipe and move down against an upflowing stream of gas ($C_p = 800$ J/kg K, $T_{g,in} = 500$ K). Plot the temperature distribution of gas and solids in the pipe, and show on the plot the outlet temperature of the two streams.

    (a) Gas flow is 10 tons/hr.

    (b) Gas flow is 20 tons/hr.

    (c) Gas flow is 8 tons/hr.

    (d) Find the mean residence time of gas $\bar{t}_g$ and of solids $\bar{t}_s$ in the contactor for the flows of part (a).

Assume perfect countercurrent plug flow contacting, fairly small particles, and a bed voidage $\varepsilon = 0.4$.

14.12. *Moving grate contactor.* Cold crushed solids (small particles, $\dot{m}_g = 10$ kg/s, $C_p = 1000$ J/kg K, $T = 300$ K) are fed to a horizontal moving grate where they are heated by hot upflowing air ($\pi = 116000$ Pa, $T = 600$ K, $v = 12$ m^3/s, $C_p = 36$ J/mol K). Estimate the temperature of the leaving solids and leaving gas for the following extremes:

    (a) if the layer of solids on the grate is thick enough so that we can reasonably assume a sharp temperature front in the vertical direction as well as in the horizontal direction.

    (b) if the solids on the grate are gently fluidized, hence well mixed in the vertical direction, but not in the horizontal direction.

14.13. Repeat problem 14.12 with one change—the flow rate of solids is $\dot{m}_s = 15$ kg/s.

14.14. Repeat problem 14.12 with one change—the flow rate of solids is $\dot{m}_s = 5$ kg/s.

14.15. To improve the efficiency of a shale processing plant fresh shale is to be preheated by hot waste combustion gases from the process. To do this a layer of finely crushed fresh shale rock (0°C) is transported along a long horizontal porous conveyor belt while the hot gas (640°C) is forced upward through the bed at a high enough velocity to just fluidize the solids.

At present the shale leaves at 480°C. What would be the outlet temperature of the solids if the gas flow rate were

    (a) raised 20%?

    (b) lowered 20%, in which case the fluidized bed collapses into a moving fixed bed of solids?

Chapter 15   HEAT REGENERATORS: DIRECT-CONTACT HEAT STORING EXCHANGERS USING A BATCH OF SOLIDS

Because solids, on a volume basis, have a very large heat capacity compared to gases, they can effectively be used as an intermediary in the transfer of heat from one gas to another. This requires a two-step operation. In the first step hot gas gives up its heat to cold solids. The solids heat up, and then in the second step the solids release this heat to a second cold gas. For continuous operations regenerators are used in pairs, as shown in Fig. 15.1.

This type of storage exchanger is used primarily when heat has to be exchanged between enormous amounts of gas, as in the steel and other metallurgical industries, or when the gases are dirty and dustladen and liable to plug up a recuperator, as is the case with flue gases in coal-burning electric power stations, or when one of the gases is too hot or reactive for the materials of construction of a recuperator, as is the case with gases from glassmaking furnaces.

Regenerators can also be designed for continuous operations as shown in Fig. 15.2.

In turn, let us take up the two major classes of regenerators: first, the fixed solid devices (the packed bed, the rotating wheel, the monolith unit), then the well-mixed solid devices (the single-stage and the multistage fluidized bed).

Fig. 15.1. A pair of cyclic regenerators for recovering heat from waste gas.

I. PACKED-BED REGENERATORS—PRELIMINARY

A. Spreading of a Temperature Front

These units are usually packed with large solids—bricks, rocks—so that the pressure drop for gas flow does not become excessive and so that fine solids suspended and entrained by the gas do not plug up the unit.

When hot gas enters an initially cold bed of solids a hot-temperature front of gas moves down the bed trailed by a hot front of solids, as shown in Fig. 15.3. Three phenomena lead to the spreading of these hot fronts:

- Deviation from plug flow of gas in the packed bed—some fluid moving faster, some moving slower. This behavior is characterized by the axial dispersion coefficient for the gas D, a sort of diffusion coefficient.
- Film resistance to heat transfer between gas and solid. Since the particles are large the interfacial area and the heat transfer coefficient ha can be very much lower than for beds of fine solids. The term ha characterizes this resistance.
- Resistance to heat flow into the particles. With large solids such as bricks and rocks the characteristic time for the heating of the particles can be large. The thermal diffusivity of the solids $k_s/\rho_s C_s$ characterizes this resistance.

Fig. 15.2. Continuously operating rotating-wheel regenerator.

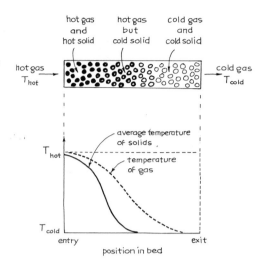

Fig. 15.3. Advancing temperature
front in gas and in solid in a packed
bed regenerator.

B. Models for the Temperature Spread

We have three levels of analysis for fixed bed regenerators, as shown in
Fig. 15.4.

1. The *flat front approximation* of Fig. 15.4a is the simplest model. It
assumes ideal plug flow of gas and immediate equalizing of gas and solid
temperature. This is a crude approximation, but still useful for baseline
estimates of behavior.

2. The *dispersion approach* of Fig. 15.4b describes each of the three
spreading factors by a diffusionlike phenomenon. This leads to a symmetrical S-shaped temperature–distance curve for solids characterized by its
variance σ^2. Assuming independence of the three spreading phenomena we
can add the variances to give

$$\sigma^2_{\text{overall}} = \underset{\text{dispersion}}{\sigma^2_{\text{gas}}} + \underset{\text{resistance}}{\sigma^2_{\text{film}}} + \underset{\text{conduction}}{\sigma^2_{\text{particle}}} \qquad (15.1)$$

This approach should reasonably approximate the real temperature distribution in a not-too-short regenerator.

3. The *rigorous analysis* which accounts properly for all three spreading phenomena should yield nonsymmetrical S-shaped curves sketched in
Fig. 15.4c. This analysis is extremely difficult and has not as yet been done.
The many approaches and partial solutions to this problem are presented
and discussed by Hausen (1983) and Jakob (1957). Other related references
are presented by McAdams (1954) and Kern (1950).

Fig. 15.4. Temperature of solids in a packed bed regenerator according to different models from the simplest to the most complicated.

We will take up the flat front and dispersion approaches in analyzing both single-pass and periodic operations of heat regenerators. We do not attempt the rigorous approach.

C. Measure of Thermal Recovery Efficiency

Suppose hot gas at $T_{h,\text{in}}$ enters a cold regenerator at T_c for a length of time t. We define the *efficiency of heat capture by solids* or the *efficiency of heat removal from the gas* for this period as

$$
\eta_h = \left(\frac{\begin{array}{c}\text{heat taken up by cold solids}\\ \text{in time } t\end{array}}{\begin{array}{c}\text{maximum possible take up}\\ \text{in time } t\end{array}} \right) = \left(\frac{\begin{array}{c}\text{heat lost by hot gas}\\ \text{in time } t\end{array}}{\begin{array}{c}\text{maximum possible}\\ \text{heat loss in time } t\end{array}} \right)
$$

$$
= \left(\frac{\overline{\Delta T_h}}{\Delta T_{\max}} \right)_{\text{of gas}} = \left(\frac{T_{h,\text{in}} - \overline{T}_{h,\text{out}}}{T_{h,\text{in}} - T_c} \right)_{\text{of gas}} \tag{15.2}
$$

Similarly for cold gas at $T_{c,\text{in}}$ entering a hot regenerator T_h for a time period t

$$\eta_c = \left(\frac{\begin{array}{c} \text{heat lost by solids} \\ \text{in time } t \end{array}}{\begin{array}{c} \text{maximum possible} \\ \text{heat loss in time } t \end{array}} \right) = \left(\frac{\begin{array}{c} \text{heat gained by cold gas} \\ \text{in time } t \end{array}}{\begin{array}{c} \text{maximum possible heat} \\ \text{gain in time } t \end{array}} \right)$$

$$= \left(\frac{\overline{\Delta T_c}}{\Delta T_{\max}} \right)_{\text{gas}} = \left(\frac{\overline{T}_{c,\text{out}} - T_{c,\text{in}}}{T_h - T_{c,\text{in}}} \right)_{\text{gas}} \tag{15.3}$$

Let us relate these efficiencies. For this consider hot fluid entering an initially cold regenerator. Making a heat balance at time t gives

$$\begin{array}{c} \text{Fraction of} \\ \text{solids heated} \end{array} = \left(\frac{\begin{array}{c} \text{heat introduced} \\ \text{by gas in time } t \end{array}}{\begin{array}{c} \text{heat needed to} \\ \text{heat up all the} \\ \text{solids} \end{array}} \right)$$

$$= \frac{\dot{m}_h C_h (T_{h,\text{in}} - T_c) t}{W_s C_s (T_{h,\text{in}} - T_c)} = \frac{\dot{m}_h C_h t}{W_s C_s}$$

The characteristic time needed to heat all the solids is then

$$\hat{t}_h = \frac{W_s C_s}{\dot{m}_h C_h} = \frac{\rho_s (1 - \varepsilon) C_s}{G_h C_h} \cdot L \tag{15.4}$$

Similarly, for the cooling of initially hot solids we have

$$\hat{t}_c = \frac{W_s C_s}{\dot{m}_c C_c} = \frac{\rho_s (1 - \varepsilon) C_s}{G_c C_c} \cdot L \tag{15.5}$$

When $\hat{t}_h = \hat{t}_c$ we have what is called *symmetric operations*. For unequal flow rates of hot and cold gas the characteristic heating and cooling times will differ, or $\hat{t}_h \neq \hat{t}_c$, and we have *unsymmetric operations*.

In periodic operations with two switching regenerators the cooling time must equal the heating time. In addition, the heat given up by hot gas must equal that gained by the cold gas. Thus, equations (15.2) to (15.5) show that the characteristic times are related to the efficiencies by

$$\frac{\eta_h}{\eta_c} = \frac{\hat{t}_h}{\hat{t}_c} \tag{15.6}$$

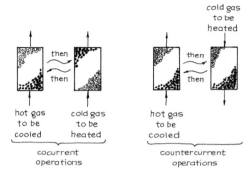

Fig. 15.5. Switching pattern for periodic cocurrent and periodic countercurrent operations of a packed bed regenerator.

D. Periodic Cocurrent and Countercurrent Operations

Periodic operations can be run in two ways. In cocurrent operations the cold fluid and the hot fluid enter one after the other at the same end of the regenerator. In countercurrent operations the hot fluid enters at one end, the cold fluid at the other end of the regenerator. These two modes are shown in Fig. 15.5.

It is not obvious which of these contacting patterns is better. The simple flat-front model says that both are equally good. However, as we will show, the dispersion model predicts that countercurrent operations has a higher efficiency. We will analyze both modes of operations.

II. PACKED-BED REGENERATORS—FLAT-FRONT MODEL

A. Cocurrent Operations with $\hat{t}_h > \hat{t}_c$

Suppose that less hot fluid flows through the regenerator than cold, and by this we mean that $\dot{m}_h C_h < \dot{m}_c C_c$ or that $\hat{t}_h > \hat{t}_c$. Then Fig. 15.6 sketches the temperature sequence in the regenerator if the switching time is taken somewhere between $\hat{t}_h$ and $\hat{t}_c$. From this sketch and equation (15.6) we find that

$$\left.\begin{array}{l} \eta_h = 100\% \\[2em] \eta_c = \dfrac{\hat{t}_c}{\hat{t}_h} \end{array}\right\} \quad \text{for } t_{sw} \text{ between } \hat{t}_h \text{ and } \hat{t}_c \text{ and for } \hat{t}_h \geq \hat{t}_c \quad (15.7)$$

Outside of this range for t_{sw} the efficiencies drop. Thus, it is best to keep t_{sw} in this range, close to $\hat{t}_h$, since this requires less frequent switching.

Fig. 15.6. Temperature front in cocurrent operations according to the flat-front model.

Looking at these efficiencies another way, although the slow-moving hot gas is able to give up all its heat, this heat is insufficient to completely heat the faster flowing cold gas.

In the special case of a symmetric regenerator, or $\hat{t}_h = \hat{t}_c$, both fluids have an efficiency of 100% (at $t_{sw} = \hat{t}$). Thus, the heat given up by the hot gas is just enough to completely heat the cold gas.

B. Countercurrent Operations with $\hat{t}_h > \hat{t}_c$; Flat-Front Model

Again, suppose that less hot fluid flows through the regenerator than cold. On sketching the moving fronts after the hot and after the cold pass we find that

$$\left.\begin{array}{l} \eta_h = 100\% \\ \\ \eta_c = \dfrac{\hat{t}_c}{\hat{t}_h} \end{array}\right\} \quad \text{for any } t_{sw} < \hat{t}_h \text{ and for } \hat{t}_h > \hat{t}_c \qquad (15.8)$$

Although we could pick any $t_{sw} < \hat{t}_h$ it is best to choose $t_{sw} = \hat{t}_h$, otherwise we would be switching more frequently than needed, or the exchanger would be larger than needed.

In the special case where $\hat{t}_h = \hat{t}_c$ (symmetric regenerator) the efficiencies are both 100% with this model as long as $t_{sw} < \hat{t}$.

C. Comments on the Flat-Front Model

The flat-front analysis shows that if the regenerator is properly operated the efficiencies of counter- and cocurrent operations are the same, and are given by equations (15.7) or (15.8). This is a useful first approximation to the behavior of the heat storing regenerator.

III. PACKED-BED REGENERATORS—DISPERSION MODEL

Consider in turn:

- the contribution of the three heat transfer resistances to the spreading of the advancing temperature front;
- the thermal efficiency of one pass operations;
- the thermal efficiency of periodic cocurrent operations; and
- the thermal efficiency of periodic countercurrent operations.

A. Evaluation of σ^2, the Quantity Which Represents the Spreading of the Temperature Front

The dispersion (or diffusion type) model leads to a symmetrical S-shaped advancing temperature front which represents the integral of the Gaussian distribution function. This S-shaped curve is characterized completely by a single quantity, the variance σ^2. Figure 15.7 shows how the temperature front spreads with σ.

From diffusion theory, if the individual contributions to the spreading act independently then one can add variances for the individual contribu-

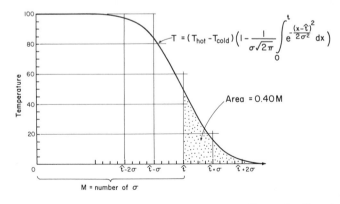

$$T = (T_{hot} - T_{cold})\left(1 - \frac{1}{\sigma\sqrt{2\pi}}\int_0^t e^{-\frac{(x-\hat{t})^2}{2\sigma^2}}dx\right)$$

Area = 0.40M

M = number of σ

Fig. 15.7. Shape of the advancing hot temperature front; according to the dispersion model.

tions, or

$$\left(\begin{array}{c}\text{Spreading of}\\\text{temperature}\\\text{front}\end{array}\right)^2 = \left(\begin{array}{c}\text{spreading caused}\\\text{by deviation from}\\\text{plug flow}\end{array}\right)^2 + \left(\begin{array}{c}\text{spreading caused}\\\text{by surface heat}\\\text{transfer resistance}\end{array}\right)^2$$

$$+ \left(\begin{array}{c}\text{spreading caused}\\\text{by resistance to}\\\text{conduction in solid}\end{array}\right)^2$$

or in symbols

$$\sigma^2 = \underset{\substack{\text{axial gas}\\\text{dispersion}}}{\sigma^2} + \underset{\substack{\text{film}\\\text{resistance}}}{\sigma^2} + \underset{\substack{\text{particle}\\\text{heating}}}{\sigma^2} \tag{15.9}$$

If it is assumed in addition that no heat travels along the solids in the direction of gas flow (reasonable for a packed bed of spherical or randomly packed nonmetallic particles, but possibly not reasonable for a compact exchanger or monolith structure), then Levenspiel (1984), using the results of Sagara *et al.* (1970), has shown that equation (15.9) becomes, term for term:

$$\frac{1}{M^2} = \frac{\sigma^2}{\hat{t}^2} = \frac{6L_p}{L} + \frac{2G_0C_g}{haL} + \frac{6}{5}\frac{G_0C_gL_p}{k_saL} \tag{15.10}$$

where

$$L_p = \frac{\text{volume of particle}}{\text{surface of particle}}, \text{ characteristic length of particle [m]}$$

$$= \frac{R}{3} \text{ for spherical particle}$$

$$= \frac{R}{2} \text{ for cylindrical particle}$$

$$a = \frac{\text{exterior surface of particles}}{\text{volume of regenerator}}, \text{ specific surface } [\text{m}^{-1}]$$

$$= \frac{3(1 - \varepsilon)}{R} \text{ for spherical particles}$$

k_s, k_g = thermal conductivity of solid and gas [W/m K]

C_s, C_g = specific heat of gas and solid [J/kg K]

$G_0 = u_0\rho$, superficial mass velocity of gas [kg/m² s]

M = number of σ between time zero and time $\hat{t}$ [−]

For spherical particles equation (15.10) becomes

$$\frac{1}{M^2} = \frac{\sigma^2}{\hat{t}^2} = \frac{d_p}{L} + \frac{1}{3(1-\varepsilon)} \cdot \frac{G_0 C_g d_p}{hL} + \frac{1}{30(1-\varepsilon)} \cdot \frac{G_0 C_g d_p^2}{k_s L}$$

(15.11)

This expression shows that as the exchanger is made longer the relative spread of the temperature front becomes smaller, hence the efficiency of the unit more closely approaches the flat-front ideal.

Let us now see how to evaluate the thermal efficiency for various operating patterns knowing the value of σ^2 from equation (15.10) or (15.11).

B. One Pass Operations; Dispersion Model

Consider a step input of hot fluid into a cold regenerator. After time $\hat{t}$ the temperature distribution of solids in the regenerator will be as shown in Fig. 15.8.

The efficiency of heat capture by the solids after time $\hat{t}$ can be found by referring to Fig. 15.8. Thus

$$\eta_{\substack{\text{single} \\ \text{pass}}} = \left(\begin{array}{c} \text{fractional} \\ \text{heat recovery} \end{array} \right) = 1 - \frac{\text{dotted area in Fig. 15.8}}{\text{area } ABCD}$$

$$= \frac{\text{shaded area in Fig. 15.8}}{\text{area } ABCD}$$

(15.12)

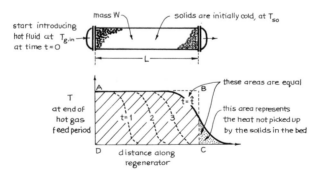

Fig. 15.8. Temperature of solids in a regenerator in a one-pass operation; according to the dispersion model.

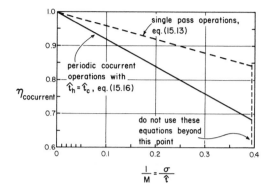

$$\frac{1}{M} = \frac{\sigma}{\hat{t}}$$

Fig. 15.9. Periodic cocurrent operations always leads to a lower efficiency than does single-pass operations. Both curves are drawn for a switching time of $\hat{t}$, and are based on the dispersion model.

Noting that the S-shaped curve of Fig. 15.8 represents the Gaussian integral shown in Fig. 15.7, and that the dotted area is calculated to be 0.4σ, the single pass efficiency is then

$$\eta = 1 - \frac{0.4\sigma}{M\sigma} = 1 - \frac{0.4}{M} \qquad \text{for } M \geq 2.5 \qquad (15.13)$$

The upper line of Fig. 15.9 then shows how the single pass efficiency depends on the value of M.

Note that the condition $M \geq 2.5$ is attached to equation (15.13). This represents operations where the solids at the inlet do reach the temperature of the hot incoming gas. This is what we call a *long regenerator*. When $M < 2.5$ the temperature profile of Fig. 15.8 is so broad that the solids at the inlet do not reach $T_{g,\text{in}}$ and the efficiency deviates from equation (15.13). We call this a *short regenerator*. From now on we only consider long regenerators.

The efficiency for an operating time $\hat{t}$ is of interest because in periodic operations a value of switching time $t_{sw} = \hat{t}$ normally is optimum in that it gives the highest efficiency of heat recovery.

C. Periodic Cocurrent Operations with Equal Flow Rates of Hot and Cold Fluid, or $\hat{t}_h = \hat{t}_c$, Dispersion Model

Suppose we choose to switch from hot to cold fluid after time $\hat{t}$, and back again after an additional time $\hat{t}$. The temperature distribution will then be somewhat as shown in Fig 15.10, and reflection shows that the in-

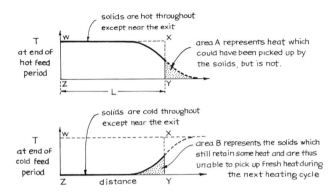

Fig. 15.10. Temperature of solids for symmetric cocurrent flow with $\hat{t}_h = \hat{t}_c = t_{sw}$; according to the dispersion model.

efficiency in heat absorption during heating is represented by the shaded areas A and B. Thus the efficiency is given by

$$\eta_{\text{heating}} = 1 - \left(\frac{\text{area } A + \text{area } B}{\text{area } WXYZ} \right)_{\text{Fig. 15.10}}$$

From equations (15.12) and (15.13) this becomes

$$\eta_{\text{heating}} = 1 - \frac{0.8}{M} \tag{15.14}$$

and from equation (15.6)

$$\eta_{\text{cooling}} = \frac{\hat{t}_h}{\hat{t}_c} \cdot \eta_{\text{heating}} = \eta_{\text{heating}} \tag{15.15}$$

An analysis of the situation leads to the following conclusions

1. The highest efficiency always occurs when the switching time t_{sw} is chosen to be $\hat{t}$.

2. Whenever $M \geq 2.5$ the succeeding temperature fronts of Fig. 15.10 do not affect each other. Thus the previous front is swept out before the next one comes along. In this situation reflection shows that the efficiency is given by

$$\eta_{\substack{\text{periodic} \\ \text{cocurrent}}} = 2\eta_{\substack{\text{single} \\ \text{pass}}} - 1 \tag{15.16}$$

3. Figure 15.9 shows how the efficiency of single-pass and periodic cocurrent operations depend on the value of M.

D. Periodic Cocurrent Operations with Unequal Flow Rates of Hot and Cold Fluid, or $\hat{t}_h \neq \hat{t}_c$; Dispersion Model

Again, assume a smaller flow rate of hot fluid than cold, or $\hat{t}_h > \hat{t}_c$, and then select t_{sw} somewhere between $\hat{t}_h$ and $\hat{t}_c$. With the flat-front assumption, Fig. 15.6 gives the temperature distributions in the regenerator. However, dispersion will give spreading fronts somewhat as shown in Fig. 15.11.

Using arguments similar to those leading to equations (15.14) and (15.15) we have, for $M > 2.5$,

$$\eta_h = 1 - \left(\frac{\text{area } H}{\text{area } WXYZ} + \frac{\text{area } C}{\text{area } W'X'Y'Z'} \right)_{\text{Fig. 15.11}} \qquad (15.17)$$

and from equation (15.6)

$$\eta_c = \frac{\hat{t}_c}{\hat{t}_h} \cdot \eta_h \qquad (15.18)$$

In evaluating these areas we first note that a switching time halfway between $\hat{t}_h$ and $\hat{t}_c$ gives close to the highest efficiency, hence close to optimum operations. Our calculation procedure for evaluating the efficiency is then as follows:

(a) Determine $\hat{t}_h$ and $\hat{t}_c$ from equations (15.4) and (15.5). Remember to choose the "hot" and "cold" fluids so that $\hat{t}_h > \hat{t}_c$.

(b) For the spreading fronts at times $\hat{t}_h$ and $\hat{t}_c$ determine σ_h and σ_c from equation (15.10).

Fig. 15.11. Temperature of solids for unsymmetric cocurrent flow with an excess of cold fluid, or $\hat{t}_h > \hat{t}_c$; according to the dispersion model.

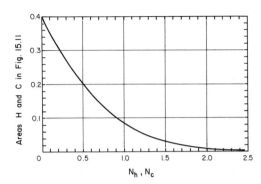

Fig. 15.12. Areas H and C from the properties of the integral of the Gaussian distribution curve.

(c) Choose $t_{sw} = \dfrac{\hat{t}_h + \hat{t}_c}{2}$.

(d) For the spreading fronts at time t_{sw} calculate

$$\sigma_{sw,h} = \sigma_h \left(t_{sw}/\hat{t}_h \right)^{1/2} \quad \text{and} \quad \sigma_{sw,c} = \sigma_c \left(t_{sw}/\hat{t}_c \right)^{1/2}$$

(e) Calculate $N_h = (t_h - t_{sw})/\sigma_{sw,h}$ and $N_c = (t_{sw} - \hat{t}_c)/\sigma_{sw,c}$, the number of standard deviations separating the $\hat{t}$'s from t_{sw}.

(f) Find areas H and C of Fig. 15.11 from Fig. 15.12.

(g) Calculate $M_h = \hat{t}_h/\sigma_{sw,h}$ and $M_c = \hat{t}_c/\sigma_{sw,c}$, the number of standard deviations between time zero and the $\hat{t}$'s.

(h) Evaluate η_h from the following form of equation (15.17):

$$\eta_h = 1 - \frac{\text{area } H}{M_h} - \frac{\text{area } C}{M_c} \tag{15.17'}$$

(i) Equation (15.6) then gives η_c.

E. Periodic Countercurrent Operations; Dispersion Model

Countercurrent flow is more difficult to visualize and analyze so let us start by considering the two extremes, first when $\hat{t}_h = \hat{t}_c$, and then when $\hat{t}_h \gg \hat{t}_c$.

1. **Extreme 1:** $\hat{t}_h = \hat{t}_c$, and $t_{sw} \leq \hat{t}$. Starting with hot fluid entering a cold regenerator, successive temperature fronts will look somewhat as shown in Fig. 15.13.

As the temperature profile shifts back and forth, from right to left, dispersion will make the front spread. However, in partially heating a cold particle it is the outside shell which heats first, and it is this which first cools

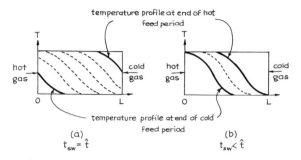

Fig. 15.13. Temperature shift in solids for symmetric counterflow, thus for $\hat{t}_h = \hat{t}_c$; according to the dispersion model.

with cold gas. This results in a self-healing of the spreading front, or a tendency to approach a flat-front profile.

The overall effect of these two opposing effects is something which cannot be evaluated by the analyses of today. Probably the best compromise at this time, and until we know better, is to assume that the temperature front remains unchanged from the one pass to the next. We use this assumption throughout for countercurrent operations.

Again, noting that the temperature profiles follow the integral of the Gaussian distribution, the thermal efficiency is evaluated with the aid of Fig. 15.14 and the following calculation procedure:

(a) Determine $\hat{t}_h = \hat{t}_c = \hat{t}$ from equations (15.4) and (15.5).
(b) Set $t_{sw} \le \hat{t}$.
(c) Determine $\sigma_h = \sigma_c = \sigma$ from equations (15.10) or (15.11).
(d) Calculate $\sigma_{sw} = \sigma(t_{sw}/\hat{t})^{1/2}$
(e) Calculate $P = (\hat{t} - t_{sw})/2\sigma_{sw}$.
(f) Calculate $Q = t_{sw}/\sigma_{sw}$.

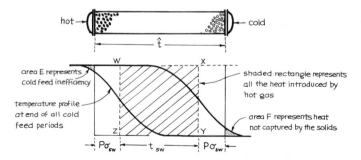

Fig. 15.14. Graph which shows how the efficiency of heat recovery is determined for symmetric counterflow, thus for $\hat{t}_h = \hat{t}_c$.

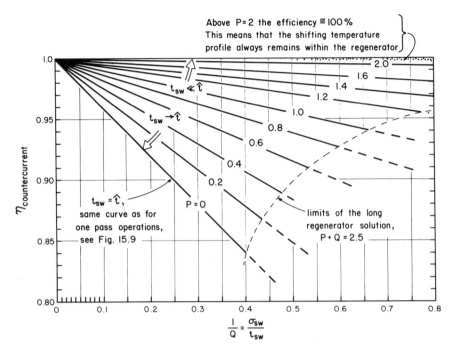

Fig. 15.15. Efficiency of symmetric counterflow, thus for $\hat{t}_h = \hat{t}_c$.

The efficiency is then given by

$$\eta = 1 - \left(\frac{\text{area } E \text{ or area } F}{\text{area } WXYZ} \right)_{\text{Fig. 15.14}}$$

This is a function of P and Q, as displayed in Fig. 15.15.

Note that if one can make $\hat{t}_{sw}$ smaller than $\hat{t} - \hat{5}\sigma_{sw}$, then one can get almost complete heat recovery, as long as the temperature profile stays balanced between the two ends of the regenerator. However, as t_{sw} is allowed to approach $\hat{t}$ the efficiency drops to that of one pass operations, and for $t_{sw} > \hat{t}$ the efficiency drops rapidly towards zero.

Comparing Fig. 15.15 with Fig. 15.9 shows that symmetric countercurrent operations are much more efficient than symmetric cocurrent operations.

2. Extreme 2: $\hat{t}_h \gg \hat{t}_c$. This situation is shown in Fig. 15.16. Here the heat from the hot gas is all captured by the solids because the whole front remains in the regenerator at the end of the hot pass (first sketch). However, this front is completely washed out of the bed by the larger flow of cold gas (second sketch). Thus the bed has no memory for the next pass of hot gas

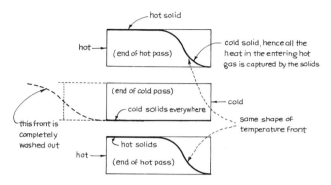

Fig. 15.16. Temperature profiles for nonsymmetric counterflow with a large excess of cold fluid, or $\hat{t}_h \gg \hat{t}_c$.

(third sketch). For this situation then

$$\eta_h = 100\% \tag{15.19}$$

$$\eta_c = \frac{\hat{t}_c}{\hat{t}_h} \tag{15.20}$$

which again is much superior to cocurrent flow.

The condition of $\hat{t}_h \gg \hat{t}_c$ is reasonably satisfied (within 5%) when

$$\hat{t}_h > \hat{t}_c + 2\sigma_c + 2\sigma_h \tag{15.21}$$

In this situation it is best to use a switching time

$$\hat{t} - 2\sigma_h < t_{sw} < \hat{t}_c + 2\sigma_c \tag{15.22}$$

This criterion also tells how long the regenerator must be (when $\hat{t}_h \neq \hat{t}_c$) so as to achieve the maximum efficiency of equations (15.19) and (15.20). The more different the ratio of the $\hat{t}_h$ to $\hat{t}_c$ the shorter the regenerator can be and still give the efficiencies of equations (15.19) and (15.20).

3. Intermediate situation. For flow ratios $\hat{t}_h/\hat{t}_c$ between these extremes we have no method today to evaluate efficiencies. All we can say is that countercurrent operations always give higher thermal efficiencies than cocurrent operations.

Fig. 15.17. Various geometries for the rotating-wheel regenerator.

F. Comments on the Dispersion Model

1. Countercurrent operations always give higher efficiencies than do cocurrent operations.

2. No matter what contacting scheme is used, the longer the exchanger, the higher will be its efficiency, with a limiting value given by the flat front model.

3. For short exchangers ($\sigma > 0.4\ \hat{t}$ or $\sigma > 0.4\ t_{sw}$) the assumptions of the dispersion model do not represent reality too well, the efficiency predictions of the model become too high, and thus the model should not be used for anything but as a crude approximation.

4. The analysis of this section only applies to exchangers having negligible heat conduction through the solids in the direction of gas flow. Thus, it applies reasonably to packed bed regenerators and to monolith regenerators made of ceramic or other poorly conducting solids.

For metallic monoliths where conduction along the metal is significant the efficiencies calculated in this chapter should be lowered somewhat, to give behavior somewhere between the packed bed and the fluidized bed (see next section).

5. For dusty gases countercurrent operations have an additional advantage in that they help avoid plugging of the regenerator.

6. For the rotating-wheel regenerator the value of the switching time is related to the time that a section of the wheel spends in the cold gas and in the hot gas. Thus, the time for one rotation of the wheel should be 2 t_{sw}.

7. For a given flow rate of hot and cold gases through a rotating regenerator changing the fraction of wheel seeing hot gas and seeing cold gas (see Fig. 15.17) does not affect operations substantially, because although this changes $t_{sw,h}$ and $t_{sw,c}$ it also changes $\hat{t}_h$ and $\hat{t}_c$ in the same proportion. Hence, it is recommended that 50% of the wheel see one fluid, 50% the other.

IV. FLUIDIZED BED REGENERATORS

Fluidized beds of fine particles are characterized by good mixing of solids and $Ua \rightarrow \infty$ (see Chap. 14). Thus, we may take the solids to be

Fig. 15.18. The heating of a batch of solids in a fluidized bed regenerator: (a) sketch of unit; (b) contacting model; and (c) temperatures in the unit.

uniform in temperature at any time, and the leaving gas to be at the temperature of the solids, as sketched in Fig. 15.18. A heat balance around the whole vessel becomes

$$\dot{m}_g C_g \left(T_{g,\,\text{in}} - T_s \right) = W_s C_s \frac{dT_s}{dt} \qquad (15.23)$$

Separating and integrating then gives

$$\frac{T_{g,\,\text{in}} - T_s}{T_{g,\,\text{in}} - t_{s0}} = \frac{\Delta T}{\Delta T_0} = e^{-t/\hat{t}} \qquad \text{where} \quad \hat{t} = \frac{W_s C_s}{\dot{m}_g C_g} \qquad (15.24)$$

Graphically this temperature progression is shown in Fig. 15.19. Note that equation (15.24) is a special case of the corresponding recuperator expression, equation (13.24), in which $Ua \to \infty$.

Let us consider the thermal efficiency of fluidized bed recuperators.

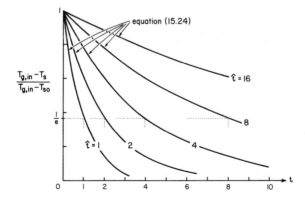

Fig. 15.19. Temperature–time curves when a batch of hot fluidized solids is cooled by a cold gas stream.

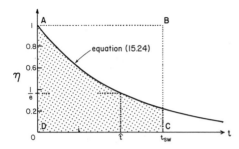

Fig. 15.20. Efficiency of heat recovery at different times during a one-pass operation of a fluidized bed regenerator.

A. Efficiency of One-Pass Operations

Suppose cold gas enters a hot fluidized bed. At the start the gas leaves hot and the efficiency is 100%. But with time the temperature of solids and exit gas will decrease as shown in Fig. 15.19 and the point efficiency of operations will likewise decrease. Referring to Fig. 15.20 the average efficiency for an operating period t_{sw} is then

$$\bar{\eta} = \frac{\text{dotted area in Fig. 15.20}}{\text{area } ABCD} = \frac{\int_0^{t_{sw}} e^{-t/\hat{t}}\, dt}{t_{sw}} \qquad (15.25)$$

Performing the integration gives

$$\bar{\eta} = \frac{\hat{t}}{t_{sw}}\left(1 - e^{-t_{sw}/\hat{t}}\right) \qquad (15.26)$$

Figure 15.21 shows that the efficiency starts at 100% at the beginning of the run and then slides to zero, thus the shorter the operating time the more efficient is the regenerator for one-pass operations.

Fig. 15.21. Average heat recovery efficiency in a one-pass operation of a fluidized bed regenerator.

B. Efficiency of Periodic Operations

First consider the situation where $\hat{t}_h = \hat{t}_c$. Here the temperature of the fluidized bed regenerator varies with time as shown in Fig. 15.22, and for optimum operations with given t_{sw} symmetry suggests that we should keep

$$1 - y_2 = y_1$$

The efficiency of operations is then obtained from integration of equation (15.25) using the limits t_1 and t_2. This gives

$$\bar{\eta} = \frac{\hat{t}}{t_{sw}}(y_1 - y_2)$$

where

$$y_1 = e^{-t_1/\hat{t}}, \qquad y_2 = e^{-t_2/\hat{t}}, \qquad \text{and} \qquad t_{sw} = t_2 - t_1$$

Combining the above five expressions, eliminating y_1, y_2, t_1, t_2, and simplifying then gives

$$\bar{\eta} = \frac{\hat{t}}{t_{sw}}\left(\frac{e^{t_{sw}/\hat{t}} - 1}{e^{t_{sw}/\hat{t}} + 1}\right) \qquad (15.27)$$

From the geometry of Fig. 15.22 we can see that when the switching time approaches zero the efficiency reaches a maximum of 50%, and as the switching time is increased the efficiency decreases towards zero.

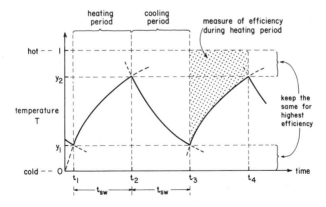

Fig. 15.22. Temperature changes for symmetric periodic operations ($\hat{t}_h = \hat{t}_c$) of a fluidized bed regenerator.

C. Comments of Fluidized Bed Regenerators

The analysis of this section leads to the following conclusions:

1. Single-stage periodic operation with $\hat{t}_h = \hat{t}_c$ has efficiencies approaching a maximum of 50% for very frequent switching times. For longer switching times the efficiency drops towards zero.

2. For periodic operations with $\hat{t}_h \neq \hat{t}_c$ similar low efficiencies are obtained.

3. Efficiencies for single-stage fluidized bed operations are very much lower than for packed bed operations, hence these units are not used in practice.

4. Multistage fluidized bed operations will give higher efficiencies than single-stage operations. However, the large power requirement needed to fluidize the large mass of solids and the still lower efficiency when compared to a well-designed packed bed regenerator are reasons for not using such units in practice.

REFERENCES

H. Hausen, *Heat Transfer in Counterflow, Parallel-Flow, and Cross-Flow* (trans. from German by M. S. Sayer), McGraw-Hill, New York (1983).

M. Jakob, *Heat Transfer*, Vol. 2, Chapter 35, Wiley, New York (1957).

D. Q. Kern, *Process Heat Transfer*, McGraw-Hill, New York (1950).

O. Levenspiel, Design of long heat regenerators by use of the dispersion model, *Chem. Eng. Sci.* **39** (1984).

W. H. McAdams, *Heat Transmission*, Third Ed., McGraw-Hill, New York (1954).

M. Sagara, P. Schneider, and J. M. Smith, The determination of heat-transfer parameters for flow in packed beds using pulse testing and chromatographic theory, *Chem. Eng. J.* **1**, 47 (1970).

EXAMPLE 15.1. THE GREAT PAPERWEIGHT DISASTER

Idiot!! Can't you read!! The order said, "One vermillion paperweight," not one million paperweights. And here we've already made 999,240 of these beauties. What will we do with the extra 999,239 that we have in our storehouse? You'd better come up with an answer quick or else you lose your job and the West Coast Paperweight Co. goes out of business.

One possibility is to sell them as packing ($\varepsilon = 0.4$) for a pair of unique, deluxe extra-high-efficiency heat regenerators, each 1 m^2 in cross section. Hot and cold gas would then be passed successively through these units at a superficial gas velocity of 4 m/s measured at 20°C and 1 atm.

(a) For cocurrent operations find the desired switching time $t_{sw} = \hat{t}$ and the efficiency of operations.

(b) For counterflow operations find the efficiency of operations for a switching time $t_{sw} = \hat{t}$.

(c) For counterflow operations find the efficiency of operations for a switching time $t_{sw} = 0.75 \, \hat{t}$.

Data: For the nearly spherical crown glass paperweights

$$d_p = 0.05 \text{ m}$$

$$\phi = 0.94, \text{ sphericity}$$

$$k_s = 1.066 \text{ W/m K}$$

$$C_s = 714 \text{ J/kg K}$$

$$\rho_s = 2500 \text{ kg/m}^3$$

Take equal quantities of hot and cold gas, and to keep things simple, assume that they have the same properties as air at 20°C. From the appendix we find these properties to be

$$\mu = 1.8 \times 10^{-5} \text{ kg/m s}$$

$$C_g = 1013 \text{ J/kg K}$$

$$\rho_g = 1.2 \text{ kg/m}^3$$

$$k_g = 0.026 \text{ W/m K}$$

$$(mw) = 0.0289 \text{ kg/mol}$$

Solution

We will solve this problem with the dispersion model, but first some preliminaries.

1. Height of paperweight-filled regenerators:

$$L = \left(\frac{999{,}238}{2 \, pw} \right) \left[\frac{\pi}{6}(0.05)^3 \frac{\text{m}^3 \text{ solid}}{pw} \right] \left(\frac{1 \text{ m}^3 \text{ reg}}{0.6 \text{ m}^3 \text{ solid}} \right) \left(\frac{1}{1 \text{ m}^2 \text{ c.s.}} \right) = 54.5 \text{ m each}$$

2. Mass flow rate and superficial mass velocity of gas:

$$\dot{m} = \left(4 \frac{\text{m}}{\text{s}} \right) \left(\frac{1 \text{ m}^3 \text{ vol}}{1 \text{ m height}} \right) \left(\frac{1 \text{ mol}}{0.0224 \text{ m}^3} \right) \left(\frac{0.0289 \text{ kg}}{\text{mol}} \right) \left(\frac{273}{293} \right) = 4.8 \text{ kg/s}$$

and

$$G_0 = \frac{\dot{m}}{A} = 4.8 \text{ kg/m}^2 \text{ s}$$

3. Heat transfer coefficient between gas and pw packing is given by equation (9.37). Evaluating the needed dimensionless groups gives

$$\text{Re}_p = \frac{d_p u_0 \rho_g}{\mu} = \frac{(0.05)(4)(1.2)}{1.8 \times 10^{-5}} = 13,333$$

$$\text{Pr} = \frac{C_g \mu}{k_g} = \frac{(1013)(1.8 \times 10^{-5})}{(0.026)} = 0.7013$$

Thus equation (9.37) becomes

$$\frac{h(0.05)}{0.026} = 2 + 1.8(13,333)^{1/2}(0.7013)^{1/3}$$

giving

$$h = 97.06 \ \text{W/m}^2 \ \text{K}$$

4. Calculate $\hat{t}$, the time needed to cool or to heat the solids. From equation (15.4):

$$\hat{t} = \frac{W_s C_s}{\dot{m}_g C_g} = \frac{[(54.5)(2500)(0.6)](714)}{(4.80)(1013)} = 12,000 \ \text{s}$$

$$= 2 \ \text{hr} \ 20 \ \text{min}$$

5. Calculate the spreading of the temperature front. From equation (15.11):

$$\frac{1}{M^2} = \frac{\sigma^2}{\hat{t}^2} = \frac{d_p}{L} + \frac{1}{3(1-\varepsilon)}\frac{G_0 C_g d_p}{hL} + \frac{1}{30(1-\varepsilon)}\frac{G_0 C_g d_p^2}{k_s L}$$

Replacing all values gives

$$\frac{1}{M^2} = 0.0009 + 0.0255 + 0.0116 = 0.0381$$

or

$$\frac{1}{M} = \frac{\sigma}{\hat{t}} = 0.1952$$

NOTE: In this situation film resistance to heat transfer contributes close to 70% to the spread of the temperature front, while heat conduction into the glass particles represents most of the remainder.

6. Single-pass efficiency. From equation (15.13) or Fig. 15.9

$$\eta_{\text{single pass}} = 1 - 0.4(0.1952) = 0.9219, \text{ or } 92\%$$

We are now ready to solve the problem.

(a) Cocurrent flow with $t_{sw} = \hat{t}$. From equation (15.16) or Fig. 15.9

$$\eta_{cocurrent} = 2(0.9219) - 1 = 0.8438, \text{ or } 84\%$$

(b) Counterflow with $t_{sw} = \hat{t}$. From the procedure leading to Fig. 15.15

$$\sigma_h = \sigma_c = \sigma_{sw} = \left(\frac{1}{M}\right)\hat{t} = 0.1952(12,000) = 2342 \text{ s}$$

$$P = \frac{\hat{t} - t_{sw}}{2\sigma_{sw}} = 0$$

$$\frac{1}{Q} = \frac{\sigma_{sw}}{t_{sw}} = \frac{2343}{12,000} = 0.1952$$

Then from Fig. 15.15

$$\eta_{counter} = 0.8438, \text{ or } 84\% \qquad \text{(same as for single pass)}$$

(c) Counterflow with $t_{sw} = 0.75$ $\hat{t} = 0.75$ (12,000) = 9000 s. Following the procedure of Section III.E we have

$$\sigma_h = \sigma_c = (0.1952)(12,000) = 234? \text{ s}$$

$$\sigma_{sw} = 2342\left(\frac{9000}{12,000}\right)^{1/2} = 2028 \text{ s}$$

$$P = \frac{12,000 - 9000}{2(2028)} = 0.7396$$

$$\frac{1}{Q} = \frac{2028}{9000} = 0.2253$$

Then from Fig. 15.15

$$\eta_{counter} \cong 97\%$$

PROBLEMS ON REGENERATORS

15.1. The geometry for the regenerator of example 15.1 is not satisfactory because it is too tall and skinny. If we made the regenerator one-quarter the height and double the diameter but maintained the same volumetric flow rate of gases, what would be the cocurrent and counterflow efficiencies of this modified design for the cases of example 15.1?

15.2. We want four times the flow rate of gases of example 15.1. For this we plan to double the diameter and reduce the height of the regenerator of example 15.1, but keep u_0 unchanged at 4 m/s. What would be the cocurrent and counterflow efficiencies of this operation for the cases of example 15.1?

In the system of example 15.1 suppose the flow rate of hot fluid is only one-half that of the cold fluid, or $u_{0,\text{hot}} = 2$ m/s and $u_{0,\text{cold}} = 4$ m/s measured at 20°C. Determine the switching time to use, and the thermal efficiency of hot and of cold fluid at this condition.

15.3. Use cocurrent flow.

15.4. Use counterflow.

15.5. Find the efficiency of cocurrent operations and the recommended switching time for the pair of regenerators of example 15.1 with just one change: the flow rate of hot fluid is 25% less than the flow rate of cold fluid, or $u_{0,\text{hot}} = 3$ m/s and $u_{0,\text{cold}} = 4$ m/s, both measured at 20°C.

   (a) First assume the flat-front approach.

   (b) Then relax this assumption and use the dispersion approach.

15.6. Repeat problem 15.5 for the case where the flow of hot fluid is just 10% less than the flow of cold fluid, or $u_{0,\text{hot}} = 3.6$ m/s and $u_{0,\text{cold}} = 4$ m/s, both measured at 20°C.

15.7. A pair of regenerators 32 m high and 3 m in diameter filled with uniformly sized and close to spherical basaltic beach stones is to be used to transfer heat from hot waste gases leaving a process to cold incoming air. For equal flows of hot and cold gas, and for the following operating conditions and properties of materials find:

   (a) the relative contributions of the three resistances to heat transfer.

   (b) the switching time to use and the efficiency of heat recovery for the best cocurrent operations.

   (c) the efficiency of heat recovery for countercurrent operations with a switching time $t_{sw} = 2$ hr.

   Data: For the solid

$$d_p = 0.08 \text{ m} \qquad k_s = 0.5 \text{ W/m K}$$

$$\rho_s = 2280 \text{ kg/m}^3 \qquad C_s = 1000 \text{ J/kg K}$$

In the regenerator

$$\varepsilon = 0.4$$

$$G_0 = 3.6 \text{ kg/m}^2 \text{ s (or } u_0 \cong 3 \text{ m/s at 20°C).}$$

For both hot and cold gas take the properties of air at 20°C and 1 atm; see example 15.1.

15.8. The laboratory's high-temperature wind tunnel experiment requires an airflow of 30 m/s at 540°C and 1 atm through a square test section 0.3 m × 0.3 m. One way to do this is to take in ambient air at 20°C and heat it continually.

(a) Find how many household electric heaters, rated at 1500 W would be needed to provide this hot air continually during the run.

Alternatively we can leisurely heat up a packed bed of rocks or other solid beforehand as shown below, and at the start of the run blow through the required ambient air at 20°C, this representing a one-pass operation.

(b) What length of packed bed and how many tons of steel balls would be required to store the heat needed for a 15-min wind tunnel experiment? Assume the flat-front approximation for this calculation.

(c) How long can we run a hot-air wind tunnel experiment with the stored heat from an 8-m-high bed of steel balls. Note that at 2.5σ away from the mean the S-shaped dispersion curve is within 1% of its asymptote.

Data: Take 300°C to be the average conditions of air in the packed bed; see diagram for additional data.

15.9. Repeat problem 15.8, but instead of using steel balls in the regenerator let us use uniformly sized crushed rock from the local rock quarry.
Data: For rock

$$d_p = 0.05 \text{ m} \qquad k_s = 0.80 \text{ W/kg K}$$

$$\rho_s = 2190 \text{ kg/m}^3 \qquad \varepsilon = 0.48$$

$$C_s = 800 \text{ J/kg K}$$

A rotary wheel regenerator is to transfer heat from hot combustion gases [6000 mol/min, 1000°C, $C_p = 30$ J/mol K, $(mw) = 0.03$ kg] to cold incoming air [6000 mol/min, 0°C, $C_p = 30$ J/mol K, $(mw) = 0.03$ kg]. The wheel is to be 1 m in diameter, 0.73 m thick, and is to consist of a tortuous path stainless-steel honeycomb ($\rho_s = 7700$ kg/m^3, $C_s = 500$ J/kg K, $\varepsilon = 0.8$). For maximum efficiency, what should be the rate of rotation of the wheel, and what can we expect the outlet gas temperatures to be?

15.10. As a first approximation:
  (1) assume no heat flow resistance from gas to metal (or $h \rightarrow \infty$) and into the metal sheet.
  (2) assume plug flow of gas through the regenerator.
  (3) assume that the regenerator is so built that no heat flows from one location in the metal to another.

15.11. Replace assumption (3) with the following:
  (3′) assume that there is no resistance to heat flow in the metal along a flow channel, thus all the metal at a given angle θ in the drawing below is at the same temperature.

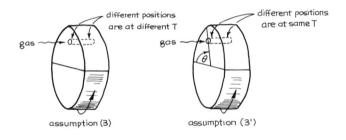

assumption (3)   assumption (3′)

15.12. A pair of identical fluidized beds is to be used to recover heat from hot waste gas leaving a process and transfer it to fresh incoming gas. What thermal recovery efficiency may be expected from this pair of regenerators if the switching time is set at
  (a) 15 min?
  (b) 30 min?
  Data: Weight of sand in each regenerator = 570 kg. For both hot and cold gas take the properties of air at 20°C. Flow rate of hot and cold gas = 0.5 kg/s.

Chapter 16   POTPOURRI OF PROBLEMS

Here are some problems which use ideas from more than one chapter.

16.1. *Quenching high-temperature reactions.* Let us use the fluidized bed to freeze a fast multistep chemical reaction and thereby obtain valuable reaction intermediates. For this introduce the hot exit gases from a jet reactor with all its reaction intermediates immediately into the bottom of a cold vigorously fluidized bed, thereby using this gas to fluidize the batch of solids, as shown in the drawing on the next page. The bed is cooled by heat transfer to the walls of a cooling section. The rest of the bed is well insulated.

Determine the temperature of the solids in the bed.

Data: For solids

$$\rho_s = 1000 \text{ kg/m}^3$$

$$d_p = 100 \text{ } \mu\text{m}$$

$$C_s = 800 \text{ J/kg K}$$

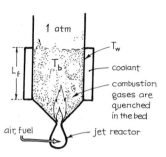

For gas:

$$C_s = 1000 \text{ J/kg K, average value}$$

$$\rho_g = 1.273 \text{ kg/m}^3 \text{, at 273 K and 1 atm}$$

$$u_0 = 0.1 \text{ m/s, measured at 273 K and 1 atm}$$

$$T_g = 1000°\text{C, entering gas}$$

For the bed:

$$d_b = 0.1 \text{ m, } \varepsilon_f = 0.48$$

Height of cooling section: $L_f = 0.4$ m

Temperature of wall of cooling section: $T_w = 0°$C

For heat loss at wall: $h_w = 191 \text{ W/m}^2 \text{ K}$

16.2. In the jet reactor of problem 16.1 heat is removed by transfer to the walls. Consider a different arrangement wherein heat is removed by passing (or flowing) solids at a steady rate through the reactor. If solids enter at 20°C and at a rate such that a vessel volume of solids is replaced every 160 s, find the temperature of gas and solids leaving the reactor
   (a) if the reactor walls are perfectly insulated;
   (b) if the reactor walls are cooled as in problem 16.1.

16.3. *Solar hot-water heaters.* Each of our horizontal rooftop solar collectors is a shallow black-painted water-filled pan about 1 m on a side, well insulated on the sides and bottom by the equivalent of 3 mm of styrofoam and covered with a thin plastic film to stop evaporation. The heated water in these units is circulated to a large heat reservoir in the basement of our home, and all works well.

   At night, however, the water in the collectors gets cold and if allowed to circulate into the basement reservoir would cool it somewhat. Hence the flow is shut off at night and the water stays stagnant in the pan all night. Determine the night temperature below which this water is likely to freeze. *Data*: The emissivity of the collector is 0.95. The temperature of the night sky in our climate is $-73°C$. For styrofoam $k = 0.035$ W/m K. Assume the worst possible conditions—still air.

16.4. *Dry scrubber for dusty gases*. Combustion Power Company uses a moving bed of granular solids as the filter medium to remove fine solids from hot gas without cooling the gas appreciably. The unit operates by having the gas move radially inward across an annulus of downflowing solids.

The solids trap the dust which is then removed on a moving grate. The clean solids are then recirculated to repeat the filtering action. If dusty gas ($C_g = 50$ J/mol K) enters at 200 mol/s, 800 K, and 120 kPa, if solids ($C_s = 1000$ J/kg K) circulate at 1 kg/s, and if the solids cool $100°C$ in the conveyor and bucket elevator, determine the temperature of the leaving gas.

16.5. *Geothermal water for heating cities*. Prodded by geologist John Hook of Salem, Oregon, Northwest Natural Gas Company is considering drilling wells on the slopes of Mt. Hood and piping this hot water ($74°C$) at 1.6 m³/s in a 1.1-m-i.d. pipe insulated with 15 mm of styrofoam to Portland, 70 km away and 760 m lower in elevation.

If too much heat is lost en route and if the water arrives colder than $65°C$, then the whole project might as well be scrapped. Calculate the temperature of the water when it reaches Portland.

(a) Account for heat loss to the surroundings only.

(b) Also include in your calculations the heat generated by friction in the pipes.

For winter conditions take the surroundings to be at $0°C$.
NOTE. You may be interested in looking at the related problems, 2.12 and 2.13.

16.6. *Plugged distributor plates.* Silane (SiH_4) is a stable gas at low temperature (below about 700 K). However, at high temperature, it decomposes spontaneously as follows

$$SiH_4(g) \xrightarrow{T > 700 \text{ K}} Si\downarrow + 2H_2$$

This is the basis for a process to produce ultrapure silicon for solar cells in which room temperature silane (300 K) passes through a porous metal distributor plate to fluidized a bed of hot silicon particles (973 K). The incoming silane gas heats up, decomposes, and the fine silicon dust formed fuses onto the particles in the bed causing them to grow.

One worry is that the entering gas may become too hot while passing through the distributor plate, forming a solid there and plugging the unit. Determine the range of superficial gas velocities of entering silane which will insure that the distributor plate never gets above 650 K, and consequently will not plug up.

Data: (i) Pressure in chamber below fluidized bed = 123 kPa. C_g (silane) = 60 J/mol K.

(ii) Between bed and distributor plate: $h = 280$ W/m^2 K.

(iii) Between distributor plate and upstream gas: $h = 40$ W/m^2 K.

(iv) It is reasonable to assume that the metal distributor plate is isothermal.

16.7. *The cooling of spent shale.* The U.S. Bureau of Mines in Albany, Oregon, is looking into a process for recovering aluminum salts from spent shale rock by water leaching. If this hot solid is dispersed directly into water, terribly noxious gases evolve and the aluminum salts are transformed into insoluble compounds. However, if the solids are first cooled before dispersing in water then no gases form and the aluminum salts remain in their soluble form, and thus may possibly be recovered economically.

Consider the cooling of 30,000 tons/day of hot spent shale from $600°C$ to $120°C$ by having them settle through a square tower penetrated by a whole host of 50-mm-o.d. horizontal tubes on 0.1-m equilateral triangle centers. These tubes are fed hot water at $100°C$ and produce steam at $100°C$. Find

(a) the cross section of tower to be used;

(b) the number of heat transfer tubes and the height of heat transfer section needed for this cooling operation; and

(c) the steam production rate.

Data: For the stream of flowing solids:

$$\overline{C}_p = 950 \text{ J/kg K} \qquad u_0 = 0.1 \text{ m/s}$$

$$\rho_{bulk} = 870 \text{ kg/m}^3 \qquad U_{to \text{ tubes}} = 150 \text{ W/m}^2 \text{ K}$$

$$\varepsilon_{bulk} = 0.46$$

hot shale ⌐⎯⎯⎤ ··· 600°C

cooled shale ··· 120°C

Also assume that the particles are small enough so that their temperature relaxation time is short enough for them to be isothermal and at the same temperature at any level in the exchanger.

16.8. Repeat the previous problem with the following changes:

1. The bare heat transfer tubes are replaced by finned tubes, which triples the surface area of the tubes, but lowers the overall U to 120 W/m^2 K, based on the outside surface area of tubes.

2. Water at 24°C and 40 MPa enters at the bottom row of the tubes and superheated steam at 400°C leaves at the top of the exchanger.

Data:

Cold water at 24°C:	$H = 100$ kJ/kg
Boiling water at 250°C:	$H = 1100$ kJ/kg
Saturated steam at 250°C:	$H = 2800$ kJ/kg
Superheated steam at 400°C:	$H = 3200$ kJ/kg

Assume that the C_p of water and steam are temperature independent.

16.9. *Silicon for solar cells*. Battelle Memorial Institute of Columbus, Ohio, is looking into a fluidized bed process for producing solar grade silicon. The reaction step for this process proceeds as follows:

$$2 \, Zn(g) + SiCl_4(g) \xrightarrow[1181 \text{ K}]{\text{above}} 2 \, ZnCl_2(g) + Si(s) \qquad \Delta H_r = -113 \text{ kJ}$$

and the reactor is to look somewhat as sketched. Now, lower temperatures give higher equilibrium conversions and are favored, but zinc condenses at 1181 K, and if this happens a fog forms which ruins operations. Thus they would like to keep the temperature above 1181 K, but as close to it as they dare.

With this in mind, Battelle plans to keep the fluidized solids at 1200 K and the reactor walls at 1185 K (for cooling purposes). A stoichiometric gas feed is to enter the 1-m high reactor at 0.15 m/s, 1200 K, and 1.1 atm. At this temperature equilibrium conversion is 72%, however the expected conversion is 60%. If $h = 160$ W/m^2 K between bed and wall calculate the largest diameter which can be used for the reactor, and find the silicon production rate (kg/hr) that can be obtained from it.

Data: For silicon: $(mw) = 0.028$ kg/gm atom.

16.10. *Oil from shale.* The United States has vast deposits of shale, a solid which contains up to about 10% organic material. There is more oil in these shale deposits than in all the petroleum reserves in the world. Crush and heat this rock and a part of this organic material comes off as volatile hydrocarbons.

The rest remains with the rock as fixed carbon which can then be burned to give off heat.

Processes designed to recover these useful volatiles face one big problem. Because so much rock accompanies each kilogram of volatiles, a significant fraction of the recoverable energy is used just to heat the rock, and is lost when the hot waste rock is discharged.

Consider the process shown in the diagram.

1. It achieves countercurrent heat exchange between the streams of hot waste solids and cold fresh solids by use of heat pipe exchangers.

2. It is able to use the not very useful fixed carbon in the shale to provide heat for the process.

3. It keeps nitrogen away from the volatiles, thus not diluting the diluting the desired product gas.

4. It uses gravity flow for all the solids.

5. It also uses gravity flow of condensate in all the heat pipes, leading to higher heat fluxes.

These are all attractive features.

For this process calculate the temperatures T_2 and T_4, and determine the fraction of what would have been wasted sensible heat which is recovered by the exchangers.

Data:

Incoming cold shale	*Leaving hot shale*
$\dot{m}_c = 2.5$ kg/s	$\dot{m}_h = 2.0$ kg/s
$C_c = 960$ J/kg K	$C_h = 900$ J/kg K
$T_1 = 0°C$	$T_3 = 800°C$

Exchanger containing the cold end of the heat pipes	*Exchanger containing the hot end of the heat pipes*
$A_c = 100$ m^2	$A_h = 80$ m^2
$U_c = 120$ W/m^2 K	$U_h = 100$ W/m^2 K

The temperature will change progressively from heat pipe to heat pipe, however one may reasonably assume that the temperature drop between the hot and the cold end of any single heat pipe is negligible.

16.11. *Heat from geothermal brine.* Hot geothermal brine contains up to 30% solids, and to cool this in an ordinary exchanger will result in horrendous scale deposits which will soon plug the unit. To overcome this problem the Idaho National Engineering Laboratory has proposed a novel exchanger design wherein the cooling brine passes through fluidized solids in the shell side of the recuperator, as sketched above. Since the total surface area of solids is so much greater than that of the tubes the salts should deposit preferentially on the particles while the tubes are kept burnished by the buffing action of the particles.

In the commercial unit brine will flow in the shell side, liquid butane in the tube side; however, in our test runs with such a unit we use water for both streams. From the data presented below determine the outside-of-tubes heat transfer coefficient h_0 for the following experimental conditions.

Data:

$$\text{Inside of tubes} \quad \begin{cases} \dot{m}' = 1 \text{ lit/s} \\ h_{i,\text{estimated}} = 3600 \text{ W/m}^2 \text{ K} \end{cases} \quad \begin{cases} T'_{\text{in}} = 5°C \\ T'_{\text{out}} = 50°C \end{cases}$$

$$\text{In the shell} \quad \begin{cases} \dot{m}'' = 1 \text{ lit/s} \\ T''_{\text{in}} = 70°C \end{cases}$$

Total area of tubes passing through the three fluidized beds: $A = 12 \text{ m}^2$. [From C. A. Allen and E. S. Grimmett, INEL report UC-66d, April 1978.]

16.12. *Freezing of people.* In certain experimental emergency procedures the anesthetized patient is plunged into ice water and his whole body is cooled to 30°C. If the heat transfer coefficient between patient and water is 32 W/m^2 K and his skin area is 2.3 m^2, estimate how long it should take to cool an 80-kg patient of medium build down to the desired state. Of course the patient is alive throughout in a sort of hibernating state, breathing slowly, heart beating and blood flowing.

(a) Assume no heat generation in the body due to metabolic action;

(b) Assume heat generation of 3200 kJ/day, or 37 W.

Clearly state any other assumptions that you need to make.

16.13. *Hot water heaters.* G. F. Montgomery of the National Bureau of Standards wrote an article entitled "Product Technology and the Consumer," in the December 1977 issue of *Scientific American*, in which he compared a conventional hot-water heater with an improved energy saving model.

(a) For the same amount of insulation around the tanks, what can you say about the difference in efficiency of the pairs of alternatives sketched in the diagram?

(b) Which of the cold water inlets (a) or (b), and which of the hot water outlets (c) or (d), represent the improved energy saving design?

16.14. *Solid to solid counterflow heat exchanger.* In a faraway gathering (IMM Conference, Sydney, Australia, 1981) O. E. Potter introduced a novel fluidized solids heat exchanger consisting of alternate counterflowing channels of hot and cold solids, somewhat as sketched in the drawing below. Consider a small two-channel test unit as shown in the drawing on the next page, each channel consisting of a three-stage fluidized bed, with a common finned wall to facilitate heat transfer between hot and cold streams. If the hot solids enter at 600°C and cold solids at 60°C find the temperature and the heat recovery efficiency of the leaving streams.

Data.

$$\text{Hot solids} \begin{cases} \dot{m}_h = 3 \text{ tons/hr} \\ C_h = 720 \text{ J/kg K} \end{cases} \qquad \text{Cold solids} \begin{cases} \dot{m}_c = 2.7 \text{ tons/hr} \\ C_c = 800 \text{ J/kg K} \end{cases}$$

side by side, multistage, fluidized beds

Area on hot side (includes all three stages and all finned surfaces) = 18 m^2. Same area on cold side. Mean heat transfer coefficient of all surfaces: $h_h = h_c = 400$ W/m^2 K. Ignore the resistance to heat transfer of the metal dividing wall and the fins.

16.15. *Heating solids with a gas.* At present a stream of solids (0°C) is heated by hot combustion gases (900°C) in two equal-sized rotating-shaft exchangers in series, in which gas and solids approximate countercurrent plug flow. Unfortunately, contacting is poor in these exchangers because gas flows over the top of most of the solids, and as a result the outlet temperatures of gas and solid are both 450°C. This is not good enough and so it has been suggested that a fluidized bed be placed between these rotating shaft exchangers, as sketched below. What size of fluidized bed would be most effective, and with this modification what would be the outlet temperature of the solids? Ignore all heat losses to the surroundings.

16.16. *Producing very pure silicon.* A number of processes have been proposed for producing ultrapure silicon from the thermal decomposition of silane

$$SiH_4(g) \xrightarrow{\text{heat}} Si\downarrow + 2H_2$$

cold
Si H₄

H₂

filter

fine silicon dust forms
in the hot reactor, 700°C

dust settles

heater melts the silicon dust

heater

1 mm droplets of Si, 1422°C

hydrogen atmosphere
100°C, 1 atm

L

quartz walls of cooling tube, 100°C

solid particles, 1402°C

In the free-space reactor concept cold gaseous silane enters a hot reactor and decomposes there to form very small particles of silicon (upper chamber of drawing). Unfortunately, the solid produced has a very large specific surface, cannot be used in this form, and must be consolidated. Hence, the settled solids are melted and are released drop by drop into a cool hydrogen-filled vertical cylinder, all kept at 100°C (lower chamber of drawing). In falling through this cylinder the particles cool and solidify.

Please come up with a rough but conservative first estimate on the length of cylinder to be used if the 1-mm silicon droplets leaving the nozzle are 10°C above their melting point, and if the particles are to be 10°C below their melting point when they reach the bottom of the cylinder.

Data: For silicon:

ρ (both solid and liquid) $\cong$ 2300 kg/m³ λ (melting) = 1.10 × 10⁶ J/kg

C_p (liquid) = 1010 J/kg K T (melting) = 1412°C

C_p (solid) = 713 J/kg K k_l and $k_s \cong$ 1 W/m K

For the quartz tube walls: T = 100°C and emissivity = 1. For a conservative estimate assume that the particles very quickly reach their terminal velocity.

Solid–solid heat exchange using a third solid. Figure 12.7 shows Shell Company's SPHER process designed to transfer heat from a stream of finely divided hot solids to a stream of finely divided cold solids. This is to be done in two fluidized beds using a stream of circulating metallic balls to soak up the heat in the lower unit and then release it in the upper unit.

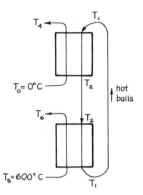

(a) Make a Q vs. T diagram to represent this system.

(b) Determine the efficiency of heat removal from the hot solids, η_h, and the efficiency of heat uptake by cold solids, η_c.

16.17. Assume plug flow of all streams.

16.18. Assume mixed flow of all streams.

16.19. Assume mixed flow of fine solids and plug flow of the large steel balls.

Data: The flow rates and properties of the flowing streams are as follows:

Hot solids	Cold solids
$T_5 = 600°C$	$T_3 = 0°C$
$\dot{m}_h = 25 \text{ kg/s}$	$\dot{m}_c = 25 \text{ kg/s}$
$C_h = 1000 \text{ J/kg K}$	$C_c = 1000 \text{ J/kg K}$
Circulating metal balls	*In the beds*
$m_b = 50 \text{ kg/s}$	$U = 200 \text{ W/m}^2 \text{ of balls} \cdot \text{K}$
$C_b = 500 \text{ J/kg K}$	$\bar{t}_{\text{fines}} = 80 \text{ s in each bed}$
$\rho_b = 5000 \text{ kg/m}^3$	$\bar{t}_{\text{balls}} = 25 \text{ s in each bed}$
$d_b = 6 \text{ mm}$	

Ignore the heat contribution of the fluidizing gases in these two exchangers, and label the temperature of the various streams as shown.

Indirect heat exchange between liquid and solid. A stream of oil is to be heated from 0°C to 400°C using the heat from a stream of hot 1000°C solids. Direct contact exchange is ruled out because the solid would contaminate the oil, so let us consider using air as an intermediary for the transfer of heat from solid to oil.

Rather finely crushed solid (C_s = 1000 J/kg K) flows at 5 kg/s, the oil (C_o = 3000 J/kg K) flows at 3 kg/s. Air (C_a = 1100 J/kg K) picks up heat from the hot solids in a direct-contact countercurrent flow raining–solids exchanger, and then delivers this heat to the oil in a countercurrent concentric pipe heat exchanger (U = 1000 W/m^2 K, A = 10 m^2). What circulation rate of air is needed to heat the oil up to the desired 400°C? Assume that no heat is lost to the surroundings from the circulating system.

It turns out that there are two solutions to this problem, one at low air flow rate, one at high air flow rate. Solve this problem

16.20. for the lower air flow rate, where

$$\phi = \frac{\dot{m}_a C_a}{\dot{m}_s C_s} < 1$$

16.21. for the higher air flow rate, where

$$\phi = \frac{\dot{m}_a C_a}{\dot{m}_s C_s} > 1$$

Label temperatures according to the flow diagram below.

16.22. In the above two problems, instead of recirculating the air why not have fresh incoming air (T_0 = 0°C) contact the solids, then the oil, and then be discarded. With this modification, all else remaining unchanged, find the minimum flow rate of air needed to heat the oil to 400°C.

16.23. *Design of an atmospheric fluidized bed combustor.* We wish to make a pre-liminary estimate of the dimensions of an atmospheric fluidized bed combus-tor for a utility scale power plant (1000 MW$_{electrical}$). The bed will consist of a fluidized mixture of coal, ash, and limestone with immersed heat transfer tubes (5-cm-o.d. on 15-cm centers) for steam generation. Boiling water enters the tubes at 204°C and steam at 538°C leaves the tubes.

(a) Find the number of 4 m × 25 m beds needed for such a plant.

(b) Find the number of rows of heat transfer tubes needed to heat the steam.

(c) Find the amount of coal needed/hr to fuel the plant.

Data:

Bed conditions: $T = 843°C$; air enters at 100°C, coal at 20°C.

C_p of all gases $= 33.4$ J/mol K

| Boiling water at 204°C: | $H = 870,000$ J/kg |

Steam at 204°C: $H = 2,795,000$ J/kg

Steam at 538°C: $H = 3,555,000$ J/kg

Assume that coal is pure carbon, that stoichiometric air is used, and that the following reaction goes to completion:

$$C + O_2 \rightarrow CO_2 \quad \Delta H_r = -393,396 \text{ J}$$

Between the tubes and the bed: $U = 250$ W/m² K. At the bed temperature the gas velocity through the bed: $u_0 = 2.5$ m/s. Efficiency of conversion of heat to steam and to electricity is 40%.

Appendix DIMENSIONS, UNITS, CONVERSIONS, PHYSICAL DATA, AND OTHER USEFUL INFORMATION

A. SI PREFIXES

Factor	Prefix name	Symbol
10^{12}	tera	T
10^{9}	giga	G
10^{6}	mega	M
10^{3}	kilo	k
10^{-3}	milli	m
10^{-6}	micro	μ
10^{-9}	nano	n
10^{-12}	pico	p

B. LENGTH

The standard unit of length is the meter.

10^{10}	10^{6}	39.370	3.2808	1	0.00062137
Å	μ	in.	ft	m	mile

2.54 cm = 1 in., 30.48 cm = 1 ft

C. VOLUME

The standard unit of volume is the cubic meter

D. MASS

The standard unit of mass is the kilogram

E. NEWTON'S LAW

$$F = \frac{ma}{g_c}$$

$$a = g = 9.80665 \text{ m/s}^2 \qquad \text{at "standard" sea level}$$

(In fact, g varies from 9.77 to 9.82 at different locations on the earth's surface.)

$$g_c = 1\frac{\text{kg m}}{\text{s}^2\,\text{N}} = 1\frac{\text{gm cm}}{\text{s}^2\,\text{dyne}} = 9.81\frac{\text{kg m}}{\text{s}^2\,\text{kg}_f}$$

$$= 32.16\frac{\text{lb}_m\,\text{ft}}{\text{s}^2\,\text{lb}_f} = 1\frac{\text{lb}_m\,\text{ft}}{\text{s}^2\,\text{poundal}} = 1\frac{\text{slug ft}}{\text{s}^2\,\text{lb}_f}$$

F. FORCE

$$(\text{Force}) = (\text{mass})(\text{acceleration})$$

the newton: $1\ \text{N} = 1\frac{\text{kg m}}{\text{s}^2} = 10^5 \text{ dyne}$

also poundal, lb_f, kg_f, gm_f

G. PRESSURE

$$(\text{Pressure}) = (\text{force})/(\text{area})$$

the pascal: $1\ Pa = 1\dfrac{N}{m^2} = 1\dfrac{kg}{m\,s^2} = 10\dfrac{dyne}{cm^2}$

$1\ atm = 760\ mm\ Hg = 14.7\ \dfrac{lb_f}{in.^2} = 29.92\ in.\ Hg = 101{,}325\ Pa \cong 34\ ft\ H_2O$

$1\ torr = 1\ mm\ Hg = 133.3\ Pa = \dfrac{1}{760}\ atm$

$1\ bar = 10^5\ Pa$ (close to 1 atm)

$1\ in.\ H_2O = 248.86\ Pa \cong 250\ Pa$

H. WORK, HEAT, AND ENERGY

$$(\text{Work}) = (\text{force})(\text{distance})$$

the joule: $1\ J = 1\ Nm = 1\dfrac{kg\,m^2}{s^2}$

10^{13}	10^6	737,562	238,846	101,972	9869	947.8	238.846	0.372506	0.277778
erg	J	ft lb$_f$	cal	kg$_f$ m	liter atm	Btu	kcal	Hp hr	kW hr

$1\ cal = 4.184\ J$

$1\ Btu = 778\ ft\ lb_f = 252\ cal = 1055\ J$

$1\ liter\ atm = 101.325\ J$

I. POWER

$$(\text{Power}) = (\text{work or energy})/(\text{time})$$

the watt: $1\ W = 1\dfrac{J}{s} = 1\dfrac{Nm}{s} = 1\dfrac{kg\,m^2}{s^3}$

$1\ kW = 239\dfrac{cal}{s} = 1.341\ Hp$

$1\ Hp = 550\dfrac{ft\,lb_f}{s} = 33{,}000\dfrac{ft\,lb_f}{min}$

J. MOLECULAR WEIGHT

In SI units, which are used throughout,

$$(mw) = \left(\frac{\text{kg}}{\text{mol}} \right)$$

$$(mw)_{O_2} = 0.032 \frac{\text{kg}}{\text{mol}}$$

$$(mw)_{\text{air}} = 0.0289 \frac{\text{kg}}{\text{mol}}, \text{ etc.}$$

K. IDEAL GAS LAW

$$pv = nRT \quad \text{or} \quad \frac{p}{\rho} = \frac{RT}{(mw)}$$

Gas constant

$$R = 8.314 \frac{\text{J}}{\text{mol K}} = 8.314 \frac{\text{m}^3 \text{ Pa}}{\text{mol K}} = 82.06 \times 10^{-6} \frac{\text{m}^3 \text{ atm}}{\text{mol K}}$$

$$= 0.08206 \frac{\text{liter atm}}{\text{mol K}} = 1.987 \frac{\text{cal}}{\text{mol K}}$$

$$= 1.987 \frac{\text{Btu}}{\text{lb mol} \,^\circ R} = 0.729 \frac{\text{ft}^3 \text{ atm}}{\text{lb mol} \,^\circ R}$$

L. DENSITY

$$(\text{Density}) = (\text{mass})/(\text{volume}), \quad \text{or } \rho = \left[\frac{\text{kg}}{\text{m}^3} \right]$$

For an ideal gas

$$\rho = \frac{p(mw)}{RT} \underset{\substack{20^\circ C, \\ 1 \text{ atm}}}{\overset{\text{air}}{=\!=\!=}} \frac{(101{,}325)(0.0289)}{(8.314)(293)} = 1.20 \frac{\text{kg}}{\text{m}^3}$$

For water

Temperature, °C	ρ, kg/m³
0–12	1000
13–18	999
19–23	998
24–27	997
28–30	996
40	992
50	988
60	983
70	978
80	972
90	965
100	958

See Appendix U for more ρ values.

M. VISCOSITY

$$\mu = \left[\frac{kg}{m\,s}\right]$$

For a Newtonian

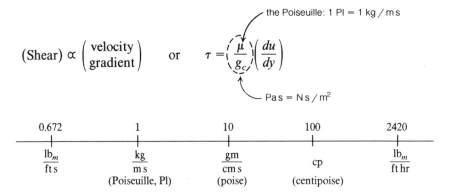

$$(\text{Shear}) \propto \left(\frac{\text{velocity}}{\text{gradient}}\right) \quad \text{or} \quad \tau = \left(\frac{\mu}{g_c}\right)\left(\frac{du}{dy}\right)$$

the Poiseuille: 1 Pl = 1 kg / m s

Pa s = N s / m²

0.672	1	10	100	2420
$\dfrac{lb_m}{ft\,s}$	$\dfrac{kg}{m\,s}$ (Poiseuille, Pl)	$\dfrac{gm}{cm\,s}$ (poise)	cp (centipoise)	$\dfrac{lb_m}{ft\,hr}$

For liquid water (20°C): $\mu = 10^{-3}$ kg/m s

For gases (20°C): $\mu \cong 10^{-5}$ kg/m s

For air (20°C): $\mu \cong 1.8 \times 10^{-5}$ kg/m s

For liquid water

Temperature, °C	μ, kg/m s
0	1.79×10^{-3}
5	1.52×10^{-3}
10	1.31×10^{-3}
15	1.14×10^{-3}
20	1.00×10^{-3}
25	0.894×10^{-3}
30	0.801×10^{-3}
40	0.656×10^{-3}
50	0.549×10^{-3}
60	0.469×10^{-3}
70	0.406×10^{-3}
80	0.357×10^{-3}
90	0.317×10^{-3}
100	0.284×10^{-3}

The following listing shows the wide range of viscosities of familiar Newtonian fluid

Fluid	μ, kg/m s
Gases	
H_2 (20°C)	0.876×10^{-5}
Steam (100°C)	1.25×10^{-5}
CO_2 (20°C)	1.48×10^{-5}
Air (0°C)	1.71×10^{-5}
(20°C)	1.83×10^{-5}
(100°C)	2.17×10^{-5}
Liquids	
Gasoline (20°C)	0.6×10^{-3}
H_2O (20°C)	1.0×10^{-3}
C_2H_5OH (20°C)	1.2×10^{-3}
Kerosene (20°C)	2.0×10^{-3}
Whole milk (0°C)	4.3×10^{-3}
Sucrose solutions (20°C)	
20 wt%	2.0×10^{-3}
40 wt%	6.2×10^{-3}
60 wt%	58×10^{-3}
70 wt%	486×10^{-3}
SAE 10W-30 motor oil (−18°C)	1.2–2.4
(99°C)	9×10^{-3}–12×10^{-3}
Olive oil (20°C)	84×10^{-3}

SAE 30W motor oil (20°C)	100×10^{-3}
Heavy machine oil (20°C)	660×10^{-3}
Glycerin (20°C)	860×10^{-3}
Molasses, very heavy (20°C)	6.6
Clover honey (20°C)	$10 - 50$
Pitch (0°C)	$5.1 - 10^{10}$

With an increase in temperature liquids become *less* viscous, gases become *more* viscous. For the viscous characteristics of non-Newtonians see Chapter 5.

N. KINEMATIC VISCOSITY

$$\nu = \frac{\mu}{\rho} = \left[\frac{m^2}{s} \right]$$

$$1 \frac{m^2}{s} = 10^4 \frac{cm^2}{s} = 10^4 \text{ stoke} = 10^6 \text{ centistoke}$$

O. THERMAL CONDUCTIVITY

$$k = \left[\frac{W}{(m^2 \text{ c.s.})(K/m \text{ length})} \right] = \left[\frac{W}{m\,K} \right]$$

$$1 \frac{W}{m\,K} = 0.00239 \frac{cal}{s\,cm\,°C} = 0.578 \frac{Btu}{hr\,ft\,°F}$$

$k \cong$ independent of pressure

For water (20°C): $k = 0.597$ W/m K

For air (20°C): $k = 0.0257$ W/m K

For steam (100°C): $k = 0.0251$ W/m K

See Appendix U for more k values

P. SPECIFIC HEAT

$$C_p = \left[\frac{J}{kg\ K} \right]$$

$$1\frac{J}{kg\ K} = 239 \times 10^{-6}\frac{cal}{gm\ °C} = 239 \times 10^{-6}\frac{Btu}{lb\ °F}$$

For water (20°C): C_p = 4184 J/kg K = 1 cal/gm °C = 1 Btu/lb °F

For air (20°C): C_p = 1013 J/kg K = 29.29 J/mol K

                         = 0.24 cal/gm °C = 7 cal/mol C

For steam (100°C): C_p = 2063 J/kg K = 37.13 J/mol K

See Appendix U for more C_p values.

Q. THERMAL DIFFUSIVITY

$$\alpha = \frac{k}{\rho C_p} = \left[\frac{m^2}{s} \right]$$

$$1\frac{m^2}{s} = 10.76\frac{ft^2}{s} = 38750\frac{ft^2}{hr}$$

For air (20°C): $\alpha = 2.12 \times 10^{-5}$ m²/s

For steam (100°C): $\alpha = 2.05 \times 10^{-5}$ m²/s

For water (20°C): $\alpha = 1.43 \times 10^{-7}$ m²/s

α values for other substances can be found from $k/\rho C_p$ values in Appendix U.

R. THERMAL RADIATIVE PROPERTIES

$$\varepsilon, \alpha = [\text{dimensionless}]$$

ε and α for room temperature radiation

$$\sigma = 5.67 \times 10^{-8} \ \text{W/m}^2 \ \text{K}^4$$

$$= 0.1713 \times 10^{-8} \ \text{Btu/hr ft}^2 \ {}^\circ\text{F}^4, \text{ the radiation constant}$$

See Chapter 12 for a short table of emissivities and absorptivities.

S. HEAT TRANSFER COEFFICIENT

$$h = \left[\frac{W}{m^2 \ K} \right]$$

1	10	10^2	10^3	10^4	10^5

Film

Forced convection, air

Condensing organics

Dropwise

Condensing steam

Free convection

Bed-wall, fluidized bed

Boiling water

Forced convection, water

$$1 \frac{W}{m^2 \ K} = 2.39 \times 10^{-5} \frac{cal}{s \ cm^2 \ {}^\circ C} = 0.1761 \frac{Btu}{hr \ ft^2 \ {}^\circ F}$$

T. DIMENSIONLESS GROUPS

Archimedes number:

$$Ar = \frac{d_p^3 \rho_g \left(|\rho_s - \rho_g| \right) g}{\mu^2}$$

Biot number:

$$Bi = \frac{hL}{k} = \frac{\left(\begin{array}{c} \text{interior resistance to heat} \\ \text{transfer in a particle} \end{array} \right)}{\left(\begin{array}{c} \text{resistance to heat transfer} \\ \text{at surface of a particle} \end{array} \right)}$$

Drag coefficient for falling particles:

$$C_D = \frac{\text{drag force, } F_d}{\left(\rho_g u^2 / 2 g_c \right) \cdot \left(\pi d_p^2 / 4 \right)}$$

Darcy friction factor for flow in pipes:

$$f_D = 4 f_f$$

Friction factor for packed beds:

$$f_f = \frac{\varepsilon^3}{1 - \varepsilon} \cdot \frac{\Sigma F g_c d_p}{u_0^2 L} = \frac{\left(\begin{array}{c} \text{frictional} \\ \text{energy loss} \end{array} \middle/ \begin{array}{c} \text{kg of} \\ \text{fluid} \end{array} \right)}{\left(\begin{array}{c} \text{kinetic} \\ \text{energy loss} \end{array} \middle/ \begin{array}{c} \text{kg of} \\ \text{fluid} \end{array} \right)}$$

Fanning friction factor for flow in pipes:

$$f_F = \frac{\tau_w}{\rho u_0^2 / 2 g_c} = \frac{\left(\begin{array}{c} \text{frictional} \\ \text{drag} \end{array} \middle/ \begin{array}{c} \text{area of} \\ \text{pipe wall} \end{array} \right)}{\left(\begin{array}{c} \text{kinetic} \\ \text{energy} \end{array} \middle/ \text{m}^3 \text{ of fluid} \right)} = \frac{\left(\begin{array}{c} \text{frictional loss} \\ \text{during flow} \end{array} \right)}{\left(\begin{array}{c} \text{kinetic energy} \\ \text{of fluid} \end{array} \right)}$$

Fourier number:

$$Fo = \frac{\alpha t}{L^2} = \frac{kt}{\rho C_p L^2}$$

Grashof number:

$$\mathrm{Gr} = \frac{L^3\rho^2 g\beta\Delta T}{\mu^2} = \frac{\text{buoyancy force}}{\text{viscous force}}$$

Hedstrom number:

$$\mathrm{He} = \frac{g_c \tau_0 d^2 \rho}{\eta^2}$$

Knudsen number:

$$\mathrm{Kn} = \left(\frac{\pi C_p}{2C_v}\right)^{1/2} \cdot \frac{\mathrm{Ma}}{\mathrm{Re}} = \frac{\left(\begin{array}{c}\text{mean free path}\\ \text{of molecules}\end{array}\right)}{\left(\begin{array}{c}\text{diameter of}\\ \text{flow channel}\end{array}\right)}$$

Mach number:

$$\mathrm{Ma} = \frac{u}{c} = \frac{\text{velocity of gas}}{\text{speed of sound}}$$

Nusselt number:

$$\mathrm{Nu} = \frac{hd}{k} = \frac{\left(\begin{array}{c}\text{total heat}\\ \text{transfer}\end{array}\right)}{\left(\begin{array}{c}\text{molecular}\\ \text{heat transfer}\end{array}\right)} = \frac{\left(\begin{array}{c}\text{conduction and}\\ \text{convection}\end{array}\right)}{\left(\begin{array}{c}\text{conduction}\\ \text{alone}\end{array}\right)}$$

Prandtl number:

$$\mathrm{Pr} = \frac{C_p\mu}{k} = \frac{\left(\begin{array}{c}\text{molecular}\\ \text{momentum transfer}\end{array}\right)}{\left(\begin{array}{c}\text{molecular}\\ \text{heat transfer}\end{array}\right)} = \frac{\left(\begin{array}{c}\text{viscous dissipation}\\ \text{of energy}\end{array}\right)}{(\text{heat conduction})}$$

$$= 0.66\text{--}0.75 \text{ for air, A, } CO_2, CH_4, CO, H_2, He, N_2$$
and other common gases
$$\cong 1.06 \text{ for steam}$$
$$= 10\text{--}1000 \text{ for most liquids}$$
$$= 0.006\text{--}0.03 \text{ for most liquid metals}$$

Reynolds number:

$$\mathrm{Re} = \frac{du\rho}{\mu} = \frac{\text{inertial force}}{\text{viscous force}}$$

U. TABLES OF PHYSICAL PROPERTIES OF MATERIALS

Solids: Metals and Alloys

| | T | k | ρ | C_p | $\alpha \times 10^6$ |
	°C	W/m K	kg/m³	J/kg K	m²/s
Aluminum	20	204.2	2707	896	84.2
Copper	20	331.9	8954	385	96.3
Iron	20	72.7	7897	452	20.4
Lead	20	34.7	11,393	130	23.4
Silver	20	406.8	10,524	235	164.5
Stainless steel	20	16.3	7820	460	4.53

Solids: Nonmetals

| | T | k | ρ | C_p | $\alpha \times 10^6$ |
	°C	W/m K	kg/m³	J/kg K	m²/s
Brick (building)	20	0.233–0.291	800–1500	—	—
Cardboard, corrugated	20	0.064	—	—	—
Chalk	50	0.9304	2000	897	0.529
Coal	20	0.1861	1400	1306	0.102
Concrete	20	1.279	2300	1130	0.492
Corkboard	20	0.0443	160	—	—
Glass	20	0.7443	2500	670	0.444
Ice	0	2.25	920	2261	1.08
Leather	30	0.1593	1000	—	—
Rubber	0	0.1628	1200	1392	0.0975
Sand	20	0.33	2600	800	0.159
Snow, dry packed	< 0	0.4652	560	2093	0.397
Wood:				—	—
Oak across grain	20	0.207	800	1759	0.147
Oak with grain	20	0.3629	800	—	—
Pine across grain	20	0.107	448	—	—
Pine with grain	20	0.2559	448	—	—

Plastics

| | k | ρ | C_p |
	W/m K	kg/m³	J/kg K
ABS	0.16–0.27	1020–1200	1510–1550
Nylon	0.17–0.34	1030–1140	1380–1670
Polycarbonate	0.19–0.22	910–1250	1170–1260
Polyethylene	0.42–0.49	913–968	2090–2300
Polyester	0.33–0.91	1100–2010	1340–2090
PVC	0.14–0.19	1240–1550	1050
Polystyrene foam	0.03–0.04	16–32	—
Polyurethane foam	0.04	122	—

Gases at 1 atm

	T °C	ρ kg/m³	C_p J/kg K	$\mu \times 10^6$ kg/m s	$k \times 10^3$ W/m K
Air	0	1.293	1005	17.2	24.4
	20	1.205	1005	18.1	25.9
	100	0.946	1009	21.9	32.1
	200	0.745	1019	25.8	39.2
	300	0.615	1046	29.5	46.0
	400	0.526	1069	33.7	52.0
	500	0.456	1093	36.2	57.4
	1000	0.277	1185	49.0	80.7
N_2	0	1.250	1030	16.7	24.3
	100	0.916	1034	20.7	31.5
	500	0.442	1105	33.9	55.8
	1000	0.268	1202	47.4	72.3
O_2	0	1.429	913	19.4	24.7
	100	1.050	934	24.1	32.9
	500	0.504	1047	40.0	61.5
	1000	0.306	1122	56.5	85.8
CO	0	1.250	1038	16.6	23.2
	100	0.916	1043	20.7	30.1
	500	0.442	1130	34.4	54.1
	1000	0.268	1231	48.7	80.6
CO_2	0	1.977	816	14.0	14.6
	100	1.447	913	18.2	22.8
	500	0.698	1156	33.9	54.9
	1000	0.423	1290	51.5	86.3
SO_2	0	2.926	607	12.1	8.4
	100	2.140	662	16.1	12.3
	500	1.033	808	31.3	30.7
	1000	0.626	867	49.2	57.6
Flue gases	0	1.295	1043	15.8	22.8
	100	0.950	1068	20.4	31.3
	500	0.457	1185	34.8	65.6
	1000	0.275	1306	48.3	109.0
H_2	0	0.0899	14070	8.4	172.1
	100	0.0657	14480	10.3	219.8
	500	0.0317	14660	16.8	387.3
	1000	0.0192	15520	23.7	571.0
NH_3	0	0.771	2043	9.4	21.0
	100	0.564	2219	13.0	34.0
	500	0.272	2918	28.1	103.6
	1000	0.165	3710	47.9	222.1
Steam	100	0.598	2135	12.0	23.7
	500	0.284	2135	28.6	68.4
	1000	0.172	2483	52.4	140.7

Liquids

	T	ρ	C_p	$\mu \times 10^6$	k
	°C	kg/m³	J/kg K	kg/m s	W/m K
Water	0	1002	4216	1792	0.5524
	20	1000	4178	1006	0.5978
	40	995	4178	654	0.6280
	60	985	4183	471	0.6513
	80	974	4195	355	0.6687
	100	961	4216	282	0.6804
	200	867	4505	139	0.6652
	300	714	5728	96	0.5396
NH_3	20	612	4798	219	0.521
CO_2	20	772	5024	70.3	0.0872
Glycerine,					
$C_3H_5(OH)_3$	20	1264	2387	1.49×10^6	0.2861
SO_2	20	1386	1365	290	0.1989
Freon-22,					
CCl_2F_2	20	1330	963	263	0.072
Ethylene glycol,					
$C_2H_4(OH)_2$	20	1116	2382	21398	0.2489
Mercury	20	13550	139	1545	7.91

SOURCES

C. P. Kothandaraman and S. Subramanyan, *Heat and Mass Transfer Desk Book*, Wiley, New York (1977).

W. H. McAdams, *Heat Transmission*, Third ed., McGraw-Hill, New York (1954).

J. H. Perry, *Chemical Engineers' Handbook*, Third ed., McGraw-Hill, New York (1950).

Plastic Desk-Top Data Bank, Cordura, 1980.

Y. S. Touloukian and co-workers, *Thermophysical Properties of Matter*, in 12 volumes, Plenum, New York (1970–   ).

AUTHOR INDEX

359

GENERAL INDEX